Introduction

Cell and Molecular Biology in Action

A series published by Addison Wesley Longman

Edited by: Dr Ed Wood, Department of Biochemistry and Molecular Biology, University of Leeds, UK

The series aims to provide introductions to key, exciting areas of cell and molecular biology, stimulating students' imaginations and initiative to bridge the gap between memorising concepts and the active approach needed for research and literature review projects. This active learning series also introduces students to experimental design and information retrieval and analysis, including exploration of the World Wide Web.

Each text in the series will cover key theory concisely and use boxes to highlight skills, techniques and applications of the theory covered. Each text will also have its own Web page providing updates and useful links to relevant sites.

For details of forthcoming titles in the series please visit the Addison Wesley Longman World Wide Web site at http://www.awl-he.com/

CELL AND MOLECULAR BIOLOGY IN ACTION SERIES

Introduction to Bioinformatics

Teresa K. Attwood and
David J. Parry-Smith

Longman

Addison Wesley Longman Limited
Edinburgh Gate, Harlow
Essex CM20 2JE
England

and Associated Companies throughout the World

© Addison Wesley Longman Limited 1999

First published 1999

ISBN 0 582 327881

British Library Cataloguing in Publication Data
A catalogue record for this book is
available from the British Library.

Library of Congress Cataloging-in-Publication Data
A catalog entry for this title is
available from the Library of Congress.

Set by 30 in Concorde BE
Produced by
Printed in Great Britain by Henry Ling Ltd., at the Dorset Press, Dorchester, Dorset.

Contents

Further information for use with this book can be accessed via a link in the catalogue entry for *Introduction to Bioinformatics* on the publisher's Web site at **http://www.awl-he.com/biology**. The aim of this facility is to provide updates and news of major advances in bioinformatics since publication. It also includes study guides for chapters, and cites additional Web site addresses which provide information of interest to biologists. We would encourage readers to use the Web site regularly.

Dedication

In special memory of Professor John and Dr Jean Utting.

Acknowledgements

We are grateful for permission to reproduce material in the following illustrations:

Figures 1.2, 9.18 and 9.19 from http://www.biochem.ucl.ac.uk/bsm/ pdbsum/ with permission from Dr R. Laskowski; Box 1.3 RUBE GOLDBERG™ and © property of Rube Goldberg, Inc. Distributed by United Media; Figure 2.4 Dr T. Etzold, European Bioinformatics Institute, Hinxton, UK; Figure 2.5 in which the maps were originally created by Dr B. Plewe for Virtual Tourist http://www.vtourist.com/webmap/maps.htm, and from http://www.angis.org.au/ with permission from Dr T. Littlejohn; Figures 3.3, 3.4, 3.8, 8.5 and 9.7 from http://www.expasy.ch/ with permission from Dr A. Bairoch; Figures 3.6, 3.7, 8.4, 9.12, 9.13 and 9.14, and Box 9.1 from http://www.blocks.fhcrc.org/ with permission from Dr J. Henikoff; Figures 3.9, 4.3, 7.2 and 9.9 from http://www.sanger.ac.uk/ with permission from Dr R. Durbin; Figure 2.3 Dr R. Lopez, European Bioinformatics Institute, Hinxton, UK; Box 8.1 and Figures 8.3, 9.10 and 9.11 from http:// www.biochem.ucl.ac.uk/cgi-bin/fingerPRINTScan/ fingerPRINTScan.cgi with permission from Mr P. Scordis, Department of Biochemistry and Molecular Biology, University College London, UK; Figure 9.15 from http://dna.stanford.edu/identify/ with permissionfrom Dr C. Nevill-Manning; Figure 9.17 from http://www.biochem.ucl.ac.uk/ bsm/cath/ with permission from Dr C. Orengo, Department of Biochemistry and Molecular Biology, University College London, UK; and Figure 10.2 from http://www.mrclmb.cam.ac.uk:80/pubseq with permission from Dr R. Staden.

We would especially like to acknowlege our gratitude to: Anne Parry-Smith for her unfailing support, and much tea and sympathy, during interminable weekends preparing the manuscript; Jeremy Packer at Cambridge Drug Discovery, Cambridge, UK for help with proof-reading; Alex Seabrook (who got us into this mess in the first place) and Kate Henderson (who helped get us out of it) at Addison Wesley Longman, whose patience (under the circumstances) has been quite remarkable, and a model to us all; and, finally, the Sequence Group in the Biochemistry Department at University College London (Phil Scordis, Julian Selley, Jane Mabey, Will Wright, and affiliate member Maria Karmirantzou) for suffering my neglect, smiling (mostly) at my bad temper, and always generously helping out whenever technology, somewhat predictably, conspired to defeat me (usually just before a deadline). Sincere thanks to you all.

Preface

The last decade has witnessed the dawn of a new era of 'silicon-based' biology, opening the door, for the first time, to the possible investigation and comparative analysis of complete genomes. In its broadest sense, genome analysis (the quest to elucidate and characterise the genes and gene products of an organism) is underpinned by a number of pivotal concepts, concerning, principally, the processes of evolution (divergence and convergence), the mechanism of protein folding, and, crucially, the manifestation of protein function.

Today, our use of computers to model such processes is limited by, and must be placed in the context of, the current limits of our understanding of these central themes. At the outset, it is important to recognise that we do not yet fully understand the rules of protein folding; we cannot invariably say that a particular sequence or fold has arisen by divergent or convergent evolution; and we cannot necessarily diagnose protein function, given knowledge only of its sequence, or of its structure, in isolation. Accepting what we *cannot* do with computers plays an essential role in forming an appreciation of what, in fact, we can do. Without this kind of understanding, it is easy to be misled, as spurious arguments are often used to promote perhaps rather overenthusiastic points of view about what particular programs and software packages can achieve.

In the field of bioinformatics, the current research drive is to be able to understand evolutionary relationships in terms of the expression of protein function. Two computational approaches have been brought to bear on the problem, tackling the identification of protein function from the perspectives of sequence analysis and of structure analysis respectively. From the point of view of sequence analysis, we are concerned with the detection of relationships between newly determined sequences and those of known function (usually within a database). This may mean pinpointing functional sites shared by disparate proteins (probably the

result of convergent evolution), or identifying related functions in similar
proteins (most commonly the result of divergent evolution).

Put like this, the identification of protein function from sequence sounds straightforward, and, indeed, sequence analysis is usually a fruitful technique; but there are two important caveats. First, in 1998, function cannot be inferred from sequence for about one third of proteins in any of the sequenced genomes, largely because biological characterisation cannot keep pace with the volume of data issuing from the genome projects (large numbers of database sequences thus either carry no annotation beyond the parent gene name, or are simply designated 'hypothetical proteins'). Second, in some instances, closely related sequences, which may be assumed to share a common structure, may not share the same function. The classic example here is lysozyme, an enzyme that catalyses the hydrolysis of bacterial cell-wall polysaccharides, which shares around 50% identity (70% similarity) with α-lactalbumin, a non-catalytic regulatory milk protein whose presence is required for lactose synthase to transfer a galactose molecule from UDPgalactose to D-glucose. In spite of this high level of sequence similarity, and highly similar folds, the functions of the proteins differ: the two key catalytic residues of lysozyme (glutamic and aspartic acid) are not conserved in α-lactalbumin; and the acidic calcium-binding motifs characteristic of α-lactalbumins are shared by only a few lysozymes. The lesson here is not that sequence or structure analysis cannot be used as a route to deducing function, but rather that *neither* technique can be applied infallibly without reference to the underlying biology.

The sequence and structure of a protein are clearly different components of its overall functionality – in mathematical terms, we might think of function as being a convolution of sequence and structure. It is perhaps useful to consider a protein fold as providing a basic scaffold, or architecture, and that this framework may be modulated, or decorated, in different ways with different sequences to confer different functions. By analogy, think of a very simple architecture – let us say an empty room. Let us now place within it a table and a chair. At this stage, we cannot tell exactly what the room is for, except perhaps that it is unlikely to be a bedroom or a bathroom. However, if we place a computer on the table, then we could say that it is more likely to be a study, for example, than a dining room. But this is not the full story. Supposing that the room might be a study has not given us an insight into the environment of the room: the study could be a room within a domestic house, where the computer is used largely, say, for keeping accounts; or it could be an office in a company or college perhaps, where the computer is used for running database searches. We only begin to get a full appreciation of the function of the room when we understand, first, how its basic framework is decorated with particular items of furniture and key accessories, and then how its location provides a context for the precise expression of its function.

At a simplistic level, we can think of proteins in much the same way. To take a trivial example, a basic fold, such as a β-barrel (our room), may occur in many different protein environments, where it may have different functions, depending on how the framework is modulated by a particular protein sequence. In this case, the folding unit may simply be the consequence of convergence to a stable architecture (a room is just a room). In the case of lysozyme and α-lactalbumin, however, which are the result of divergent evolution, both sequence and structure are similar (our room has essentially the same furniture). Here, the different functions of the proteins can only be resolved at the level of *specific* residues that mediate their different activities (i.e., the room's accessories differ – e.g., the computer is a powerful workstation rather than a PC).

The message here is that Nature has her own complex rules, which we only poorly understand, and which we cannot easily encapsulate within computer programs. No current algorithm can 'do' biology. Programs provide mathematical, and therefore infallible, models of biological systems. To interpret correctly whether sequences or structures are meaningfully similar, whether they have arisen by the processes of divergence or of convergence, whether similar sequences, or similar folds, have the same or different functions: these are challenging problems, even for the experienced biologist. There are no simple solutions, and computers do not give us the answers; rather, given a sea of data, they help to narrow the options down so that we, the users, can begin to draw informed, biologically reasonable conclusions.

The issues relating to sequence and structure analysis are many and varied, and separately they constitute vast disciplines. Although, perhaps ideally, investigation of protein function should draw on insights derived from both sequence and structure, we will see that, at the present time, the amount of available structural information is limited by comparison with the quantity of available sequence data. It is clear that the three-dimensional structure of proteins is more faithfully preserved, or conserved, than is the underlying sequence. Thus, seemingly different sequences may adopt remarkably similar folds; in studying evolutionary relationships, this presents the sequence analyst with some daunting challenges. In light of the tidal wave of genome data now before us, these challenges, far from diminishing, are growing in scale. This book therefore considers the problems of detecting the distant relationships sequestered in these oceans of data solely from the perspective of sequence analysis, and, in this context, explores both the possibilities and the current limitations of biology *in silico*.

Overview

Following our opening thoughts, let us now try to make two things clear: first, what this book is *not* about, and second, how what it *is* about is structured, in order to give a better understanding of how to use the book and what to expect from it.

So, to begin with, this text is not about protein structure, structure analysis or structure prediction (secondary or tertiary); there are already numerous excellent and authoritative books in this area. Similarly, this is not a biology text, although we stress at all times that results of sequence analysis, and especially functional inference, must be placed in a proper biological context (and, in the limit, can only be verified by direct experiment).

This book is about sequence analysis. It is an attempt to discuss what realistically can and cannot be achieved with today's computer programs in today's databases. Sequence analysis does not give black and white answers regarding 3D molecular structure and function, and/or evolutionary relationships. Computational methods merely provide clues; the challenge is to design analysis strategies that most effectively capture known biological knowledge and hence can offer insights that might ultimately suggest particular experiments. Thus, wisely used, sequence analysis is a valuable tool in the trade of modern molecular biology.

The birth of genome projects, and the accompanying deluge of sequence information, has placed sequence analysis firmly in the spotlight, as computational methods are now needed to try to infer biological information from sequence data alone. The analysis approaches discussed in this book are concerned with the ability to detect relationships with sufficient confidence that structural or functional information about a known sequence may be sensibly transferred to an uncharacterised one. Such relationships may be quite distant, and similarities between them confined to only small regions. Discovering such islands of similarity between ancient evolutionary relatives within the ocean of noise that biological databases contain is a major signal-to-noise issue. Consequently, many different techniques have been designed to tackle this problem from a variety of different perspectives; to the newcomer, these may seem rather bewildering. The aim of this book is not to focus on any analysis method in particular, but rather to provide a guide to some of the most popular databases and search tools currently available. Hence, we encourage the would-be sequence analyst not to rely on specific methods, but to explore different databases with different tools, and to try to establish a consensus view from the various approaches.

The reasons for taking this stance are pragmatic: e.g., no database is yet complete, so there is no 'single' database for the job; contents of similar resources, in some cases, only partially overlap, so to omit one is possibly to compromise the effectiveness of a given search; no database search or sequence alignment tool is infallible, so different methods can usefully provide a 'reality check' on one's results; some databases and pattern recognition techniques are still very much at the research level and have not stood the test of time – their availability is therefore not always guaranteed (because, for example, financial support for a particular research project has ended). The flavour of the book is therefore deliberately philosophical in places, in an attempt to place emphasis on underlying *concepts*, knowing that, in this rapidly developing area, the specifics will change very quickly.

Nevertheless, clearly many important methods and databases are required as part of the sequence analyst's armoury. The book therefore begins by outlining, chapter by chapter, some of today's most commonly used databases, information resources and analysis methods. Although perhaps tedious to encounter in quite this form, the essence of the approach is to provide sufficient familiarity with the background material to be able to begin hands-on practical sequence analysis. Building on the theory at the heart of the book, we therefore delve into a real application, in the form of an interactive bioinformatics practical on the World Wide Web. The book could therefore be regarded as a detailed manual that accompanies the Web tutorial.

Figure i illustrates how the book is structured to pave the way to the practical in Chapter 9; the contents of individual chapters are outlined below. Broadly, there are four main themes: first, where and what are the databases (protein and nucleic acid sequence), and what are their formats? Second, what are the sequence alignment methods (pairwise and multiple), and what are their strengths and weaknesses? Third, what are the pattern recognition methods (single- and multiple-motif, profile, etc.), and what are the pitfalls? And, finally, how do we combine these techniques within an effective search protocol?

1. Introduction

We begin at a basic level, exploring such questions as 'What is bioinformatics?' and 'Why is it important?' To set these questions in context, some of the historic developments that led from labour-intensive manual protein sequencing to the current information revolution are outlined (where once

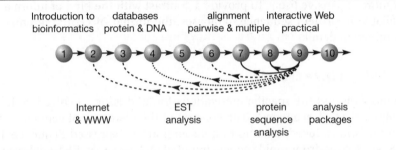

Figure i Book overview, illustrating the main themes of successive chapters, which pave the way to an interactive Web practical. Although aspects of all chapters ultimately bear on the practical, they do not all carry the same weight – this difference of emphasis is highlighted by means of the relative weights of the lines linking back to the referring chapters.

it took years to determine a complete protein sequence, new conceptual translations are entering our databases at the now-alarming rate of one per minute!). The ramifications of the biological sequence/structure deficit, and the consequent imperative to analyse sequences as a route to deducing protein function, are discussed. The feasibility and status of protein secondary and tertiary structure prediction are also briefly mentioned. Some fundamental definitions (e.g., homology vs. similarity) are given.

2. Information networks

The Internet, the World Wide Web, and the global network of biological information and service providers are introduced. We refer to the browsers that have made such information and facilities accessible, and to the software tools that have been developed for linking and searching distributed databases. Web addresses of some important bioinformatics organisations are summarised.

3. Protein information resources

Now we focus on some of the most important protein databases accessible via the Internet and the Web. The different levels of stored data (e.g., primary, secondary, tertiary) are discussed, and formats of the most widely used primary and secondary resources (e.g., SWISS-PROT and PROSITE) are explained in detail. Reasons for the evolution of composite databases and of integrated database projects are also briefly mentioned.

4. Genome information resources

Here, we take a closer look at DNA sequence data repositories, including the primary producers (GenBank, EMBL, DDBJ) and a range of specialist genome information resources (for obvious reasons, we can only hint at the range of databases currently available, and readers are referred to the Web for more expansive lists). To provide a contrast with the kind of information available in protein sequence databases, the format of GenBank entries is examined in detail.

5. DNA sequence analysis

We now discuss the specific motivations for, and issues involved with, the analysis of DNA sequences. The concept of the hierarchy of genomic information is introduced, leading to a discussion of Expressed Sequence Tags (ESTs), derived from rapid sequencing of cDNA libraries. EST analysis provides the focus for this chapter, largely because of its growing importance in the contexts of gene- and drug target-discovery programmes. The particular problems inherent in EST sequence analysis are therefore described, and a practical example is discussed. Three producers of EST data are profiled.

6. Pairwise alignment techniques

Pairwise comparison is a fundamental task in sequence analysis, providing the basis of database search algorithms, which seek to determine whether sequences are significantly similar, and hence whether or not they are likely to be homologous. The concepts of identity and similarity are described, and we consider the definition of local and global similarity, examining in some detail the algorithms that fall into these two categories.

7. Multiple sequence alignment

Seeking relationships between pairs of sequences (your query, say, and a database sequence) is only a first step in the analysis process. Often, interest hinges on groups of sequences that form *gene families*; in these cases, it is desirable to be able to trace the connections within such groups, in order to identify conserved family characteristics. The process of multiple alignment effectively enhances the signal-to-noise ratio within sets of sequences, ultimately facilitating the elucidation of biologically significant motifs (which may be diagnostic of structure or function). Here, we review some of the different approaches to multiple alignment, including fully manual and automatic techniques.

8. Secondary database searching

Building on the themes of primary database searching (pairwise alignment) and of multiple sequence alignment, we now review the analysis methods that underpin secondary database searches. The format of secondary databases was introduced in Chapter 3. Here, attention is turned to the types of information stored in the major resources, including regular expressions, profiles, fingerprints, blocks and Hidden Markov Models. These approaches essentially use multiple alignments to characterise protein families in different ways, often with notably different levels of success. The main diagnostic strengths and weaknesses of the techniques are therefore highlighted. We advise that, in view of the fallibility of the different pattern recognition methods, and given that the databases do not entirely overlap in content, good analysis strategies should sensibly include them all.

9. Building a sequence search protocol

Now armed with the concepts of primary and secondary database searching, we consider how to bring these ideas together within a generalised sequence analysis protocol, with particular reference to an interactive bioinformatics practical on the World Wide Web. The idea is to learn how to interpret results of searching the different data types (inevitably, search outputs are quite different – some are more opaque and difficult to fathom than others!), and to appreciate the difference between biological and mathematical significance – when, for example, is a hit not a hit? With these issues in mind, the chapter is intimately linked with a Web tutorial, which seeks to identify an unknown fragment of DNA via hands-on use of the primary, secondary and structure classification databases. The practical offers a step-by-step guide through a particular analysis protocol, with embedded help, explanatory diagrams, and further information pages, which themselves link back to this text. Once again, the principle is to provide general concepts and hence to provide the reader with confidence to devise his or her own search strategy.

10. Analysis packages

Having discussed the seemingly bewildering array of biological databases, the algorithms for searching them, and the kind of analysis strategy that may be pursued via a simple Web interface, we briefly examine stand-alone analysis suites. Popular packages from the commercial and public sectors are reviewed (e.g., GCG, Staden), and the latest developments on the WWW are highlighted (e.g., CINEMA); we touch on the types of facilities offered, and consider why such packages have evolved, and what the future might hold. Once again, in a book of this sort, we can only give a flavour of the kinds of package currently available, and readers are referred to the Web for more comprehensive lists. Issues relating to licensing are mentioned from both academic and commercial perspectives.

In light of the growing body of jargon associated with this field, a glossary is provided at the end of the book; this will hopefully reduce frustrating, and possibly fruitless, back-tracking to find prior definitions of seemingly unexplained words, phrases and abbreviations.

Finally, a few words to keep in mind while reading this book – some ground rules if you like: (i) don't always believe what databases tell you (the information they contain may be misleading and is sometimes wrong – recent estimates of genomic sequencing errors have been in the range

0.1–4.0% of nucleotides, affecting more than 5% of proteins, and while errors of annotation cannot sensibly be quantified, some scientists fear that current automated methods for function assignment may be leading us to a future error catastrophe); (ii) don't always believe what programs tell you (the results they provide may be misleading and are sometimes wrong – computer programmers do occasionally make mistakes); (iii) don't always believe what Web servers tell you (the results they return may be misleading and are sometimes wrong – authors of Web interfaces, even at the most prominent of bioinformatics centres, also occasionally make mistakes, as we discovered while writing this book); and (iv) don't always believe everything you read – mistakes abound, even in the published literature (and even in some of the classic articles cited herein). In short, don't be a naïve user, but think about, question and, above all, be critical of the information you gather. Where possible, try to make sense of the whole picture, rather than grasping at more palatable tiny parts. Only when a pattern begins to emerge from the noise, and starts to form coherent threads, can you begin to be confident that your analysis is moving in the right direction and that your conclusions might in fact be credible.

CHAPTER ONE

Introduction

1.1 Introduction

The purpose of this chapter is to describe what is meant by bioinformatics and to explain its importance in modern molecular biology. We begin by outlining some of the historic developments that led from labour-intensive manual protein sequencing, to today's information deluge arising from the development of fully automated DNA sequencing technologies. The chapter continues with a description of the biological sequence/structure deficit, and the consequent imperative to analyse sequences as a route to deducing protein function – in this context, the current status of protein structure prediction is also briefly mentioned. Important questions concerning the feasibility of protein tertiary structure prediction are framed, with considerations of how much of the information for folding resides in the primary structure, and the nature of the role of molecular chaperones. Reminders of the definitions of primary, secondary, tertiary and quaternary structure are given. Within the text, here and in all subsequent chapters, terms first rendered in **bold** are defined in the Glossary.

1.2 The dawn of sequencing

1.2.1 Protein sequencing

The science of **sequencing** began slowly. The earliest techniques were based on methods for separation of **proteins** and peptides, coupled with methods for identification and quantification of **amino acids**. Prior to 1945, there was not a single quantitative analysis available for any one protein. However, significant progress with chromatographic and labelling techniques over the next decade eventually led to the elucidation of the first complete sequence, that of the peptide hormone insulin (Ryle *et al.*, 1955). Yet it was a further *five* years before the sequence of the first **enzyme** was complete – this was ribonuclease (Hirs *et al.*, 1960). By 1965, around 20 proteins with more than 100 residues had been sequenced, and by 1980 the number was estimated to be of the order of 1500. Today, with more than 300 000 sequences available, it is hard to imagine such a slow awakening.

Initially, the majority of protein sequences were obtained by the manual process of sequential **Edman degradation – dansylation** (Edman, 1950). A key step towards the rapid increase in the number of sequenced proteins was the development of automated sequencers, which, by 1980, offered a 10^4-fold increase in sensitivity relative to the automated procedure implemented by Edman and Begg in 1967.

Advances in mass spectrometry also allowed significant progress, the first complete protein sequence assignment by this method being achieved in 1979. Mass spectrometry has the particular advantage that it can identify **post-translational modifications**, which may be lost with other analytical approaches. Thus, for example, the technique played a vital role in the discovery of the amino acid γ-carboxyglutamic acid, and its location in the N-terminal region of prothrombin.

1.2.2 Nucleic acid sequencing

In the 1960s and 1970s, scientists struggled to develop methods to sequence nucleic acids, but the first techniques to emerge were really only applicable to **RNA (ribonucleic acid)**, especially transfer-RNAs (tRNA). tRNAs were ideal subjects for this early work, first, because they are short (typically 74–95 **nucleotides** in length), and second, because it is possible, if not easy, to purify individual molecules.

DNA (deoxyribonucleic acid) is not like this. Human **chromosomal** DNA molecules may contain between 55×10^6 and 250×10^6 **basepairs (bp)**, orders of magnitude larger than RNAs. Assembling the complete nucleotide sequence of an entire chromosomal DNA molecule is a huge task. Even if the sequence can be broken into smaller components, purification remains a problem. The longest fragment that can be sequenced in one experiment is ~500 bp. Analysis of a human chromosome could therefore yield ~0.5×10^6 fragments. How then can a single fragment be separated from all others?

The advent of **gene cloning** and PCR provided the solution. With these methods, it became possible to purify defined fragments of chromosomal DNA, thus paving the way for the emergence of fast, efficient DNA sequencing techniques. By 1977, two sequencing methods had emerged, using chain termination and chemical degradation approaches. With only minor changes, the techniques propagated to laboratories throughout the world, and laid the foundation for the sequence revolution of the 1980s and 1990s, and the subsequent birth of bioinformatics.

1.3 What is bioinformatics?

During the last decade, molecular biology has witnessed an information revolution as a result both of the development of rapid **DNA sequencing** techniques and of the corresponding progress in computer-based technologies, which are allowing us to cope with this information deluge, in increasingly efficient ways. The broad term that was coined in the mid-1980s to encompass computer applications in biological sciences is **bioinformatics**.

The term bioinformatics has been commandeered by several different disciplines to mean rather different things. In its broadest sense, the term can be considered to mean information technology applied to the management and analysis of biological data; this has implications in diverse areas, ranging from artificial intelligence and robotics to **genome** analysis. In the context of genome initiatives, the term was originally applied to the computational manipulation and analysis of biological *sequence* data (DNA and/or protein). However, in view of the recent rapid accumulation of available protein structures, the term now tends also to be used to embrace the manipulation and analysis of three-dimensional (3D) structural data.

1.4 The biological sequence/structure deficit

It is instructive to bear in mind the difference of scale in handling sequence and structural information. At the beginning of 1998, in publicly available, non-redundant databases, more than 300 000 protein sequences have been deposited, and the number of partial sequences in public (Boguski *et al.*, 1994) and proprietary **Expressed Sequence Tag (EST)** databases (see Chapter 4) is estimated to run into millions. By contrast, the number of unique 3D structures in the Protein DataBank (PDB) (Bernstein *et al.*, 1977) is still less than 1500. Although structural information is far more complex to derive, store and manipulate than are sequence data, these figures nevertheless highlight an enormous information deficit (see Figure 1.1); this situation

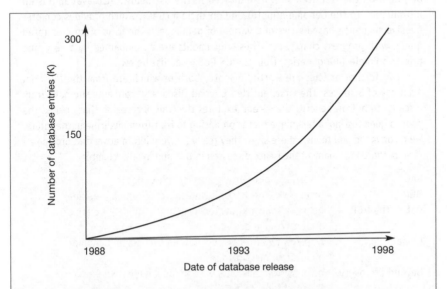

Figure 1.1 The protein sequence/structure deficit in 1998. The graph illustrates the non-redundant growth of sequence data during the last decade (—) and the corresponding growth in the number of unique structures (—).

is likely to get worse as the various **Genome Projects** around the world begin to bear fruit. Of course, the acquisition of structural data is also hastening, and future large-scale structure determination enterprises could conceivably furnish 2000 3D structures annually; but this is a small yield by comparison with that of sequence databases, which are doubling in size every year, with a new sequence being added, on average, once a minute!

1.5 Genome projects

In the mid-1980s, the United States Department of Energy (DoE) initiated a number of projects to construct detailed genetic and physical maps of the human genome, to determine its complete nucleotide sequence, and to localise its estimated 100 000 **genes.** Work on this scale required the development of new computational methods for analysing genetic map and DNA sequence data, and demanded the design of new techniques and instrumentation for detecting and analysing DNA. To benefit the public most effectively, the projects also necessitated the use of advanced means of information dissemination in order to make the results available as rapidly as possible to scientists and physicians. The international effort arising from this vast initiative became known as the Human Genome Project (see Box 1.1).

BOX 1.1: GENOMES AND THE PATH TO HUMAN BENEFIT

An excellent guide to the world-wide Human Genome Project (HGP) and a review of the role, history and achievements of the US Department of Energy in the HGP can be found on the HGP Web site, as indicated in the list below. Also available is an introduction to the Genome Annotation Consortium (GAC), which provides comprehensive sequence-based views of a variety of genomes in the form of an illustrated guide, with progress charts, etc. These documents are accompanied by a very fine primer on molecular genetics, from the US DoE – see list below.

Two further articles are worthy of note, both of which are from the National Academy of Sciences: the first, entitled Beyond Discovery, concerns the path from research to human benefit; the second explores the trail of research that opened the door to gene testing, a technique that is promising to transform medicine in the future. We recommend you to visit these sites; they really are very informative, and provide an ideal backdrop for many of the issues discussed in this and in later chapters.

GAC	http://compbio.ornl.gov/gac/index.shtml
HGP	http://www.ornl.gov/TechResources/Human_Genome/
DoE in the HGP	http://www.ornl.gov/TechResources/Human_Genome/ publicat/tko/index.htm
Primer	http://www.ornl.gov/TechResources/Human_Genome/ publicat/primer/intro.html
Beyond Discovery	http://www4.nas.edu/beyond/beyonddiscovery.nsf/ frameset?openform
Gene Testing	http://www4.nas.edu/beyond/beyonddiscovery.nsf/ DocumentFrameset?OpenForm&HumanGeneTestin

Similar research efforts were also launched to map and sequence the genomes of a variety of organisms used extensively in research laboratories as **model systems**: these included the bacterium *Escherichia coli,* the yeast *Saccharomyces cerevisiae,* the nematode worm *Caenorhabditis elegans,* the fruit fly *Drosophila melanogaster,* the common weed *Arabidopsis thalania,* and the domestic dog *Canis familiaris* and mouse *Mus musculus.* In April 1998, although the sequencing projects of only a small number of relatively small genomes had been completed (see Table 1.1), and the human genome is not expected to be complete until after the year 2000, the results of such projects were already beginning to pour into the public sequence databases in overwhelming numbers.

Table 1.1 Completed genomes in April 1998: there is only one complete eukaryotic genome (*S. cerevisiae*), with 17 more in progress; and more than 20 complete prokaryotic genomes, with 45 in progress. For more information see **http://www-fp.mcs.anl.gov/~gaasterland/genomes.html**

Organism	Laboratory	Genome size (megabases)
Saccharomyces cerevisiae	Europe	16.0 (Goffeau et al., 1996)
Pseudomonas aeruginosa	USA	5.9
Escherichia coli	USA/Japan	4.6
Bacillus subtilis	Europe/Japan	4.2
Synechocystis sp. PCC6803	Japan	3.5
Bacillus sp. C-125	Japan	4.2
Streptococcus pneumoniae	USA	2.5
Neisseria meningitidis	Europe	2.2
Archaeoglobus fulgidus	USA	2.2
Neisseria gonorrhea	USA	2.2
Pyrococcus furiosus	USA	2.1
Pyrococcus horikoshii	Japan	2.0
Pyrobaculum aerophilum	USA	1.9
Haemophilus influenzae	USA	1.8 (Fleischmann et al., 1995)
Methanobacterium thermo.	USA	1.8
Methanococcus jannaschii	USA	1.8 (Bult et al., 1996)
Streptococcus pyogenes	USA	1.8
Helicobacter pylori	USA	1.7
Aquifex aeolicus	USA/Europe	1.5
Treponema pallidum	USA	1.1
Rickettsia prowazekii	Europe	1.1
Borellia burgdorferi	USA	1.0
Mycoplasma pneumoniae	Europe	0.8
Ureaplasma urealyticum	USA	0.8
Mycoplasma genitalium	USA	0.6 (Fraser et al., 1995)

1.6 Status of the human genome project

Up to about mid-1998, the vast publicly funded efforts to sequence the human genome were predicting that sequencing would probably not be complete until ~2003–2005. To date, a two-step process has been used to sequence the genome by analysing mapped clones covering the human chromosomes. The first step has been so-called '**shotgun**' sequencing and assembly of random fragments from each clone; this is followed by an expensive, labour-intensive 'finishing' process, involving the closing of gaps and resolution of ambiguities.

In May 1998, estimates for completing the genome were shaken by the announcement that the president of The Institute for Genomic Research in Rockville, Maryland, a leading figure in the field of genome sequencing, is to form a company with Perkin-Elmer Corp.; the company plans to sequence the genome in three years, using a whole-genome shotgun approach. This venture involves breaking the genome into random unmapped fragments for sequencing. But there is doubt whether the entire genome, 70% of which consists of highly repetitive sequences, can in fact be reassembled.

Uncertainties surrounding this strategy, and fears of data-hoarding by a private venture, inspired a renewed thrust to step up public efforts. The public players thus proposed to create a 'rough draft' of the genome within three years, which would be ~95% complete. The new approach involves dramatic acceleration of the shotgun phase, which is the simplest and cheapest (costing ~10 cents per base), and could provide a high-quality draft by 2001.

The rough draft is not intended to be a substitute for the complete, highly accurate finished sequence, but rather an intermediate product. Although there are fears that producing the draft will distract sequencers away from the primary goal (i.e., the finished whole), it is nevertheless felt that it will allow other scientists to proceed more rapidly with projects that exploit newly determined sequence data (e.g., from discovering disease genes, to molecular characterisation of disease genes that have been mapped but not yet identified). Whatever the eventual completion date, the actual outcome will clearly be *more* sequence data, *sooner*.

1.7 Why is bioinformatics important?

In a field that has been dominated by structural biology for the last 20–30 years, we are now witnessing a dramatic change of focus towards sequence analysis, spurred on by the advent of the genome projects and the resultant sequence/structure deficit. The central challenge of bioinformatics is the rationalisation of the mass of sequence information, with a view not only to deriving more efficient means of data storage, but also to designing more incisive analysis tools. The imperative that drives this analytical process is the need to convert sequence information into biochemical and biophysical knowledge; to decipher the structural, functional and evolutionary clues encoded in the language of biological sequences.

It is clear that mere acquisition of sequences conveys little more about the intricate biology of the systems from which they are derived than a company phone directory can reveal about the complexities of the company's business. To extract biological meaning from sequence information is an exacting science. In essence, we are faced with the task of decoding an unknown language. This language may be decomposed into sentences (proteins), words (motifs), and letters – its alphabet – (amino acids), and the code may be tackled at a variety of these levels. By themselves, the letters have no higher meaning, but their particular combination into words is important. Sometimes, the most subtle of changes, a single letter within a word perhaps, can change its meaning (e.g., hog – hag), and hence the meaning of the entire sentence; so it is vital to decipher the code correctly. Consider, for example, the single base change in the human haemoglobin A chain codon for glutamic acid (GAA) to valine (GUA); in homozygous individuals, this minute difference results in a change from a normal healthy state to fatal sickle cell anaemia.

Ultimately, our aim is to be able to understand the words in a sequence sentence that form a particular protein structure, and perhaps one day to be able to write sentences (design proteins) of our own. Today, application of computational methods allows us to recognise words that form characteristic patterns or signatures, but we do not yet understand the intricate syntax required to piece the patterns together and build complete protein structures.

In investigating the meaning of sequences, two distinct analytical themes have emerged: in the first approach, pattern recognition techniques are used to detect similarity between sequences and hence to infer related structures and functions; in the second, *ab initio* prediction methods are used to deduce 3D structure, and ultimately to infer function, directly from the linear sequence. The development of more powerful pattern recognition and structure prediction techniques will continue to be dominant themes in bioinformatics research while the number of experimentally determined protein structures remains small.

1.8 Pattern recognition and prediction

At the outset, it is important to highlight the distinction between pattern recognition and prediction. As mentioned above, these are the principal analytical approaches in bioinformatics, and the concepts are often used interchangeably. However, in terms both of what they attempt to achieve and of what, in fact, they can achieve, these methods are really quite different and should not be confused.

Pattern recognition methods, as the name suggests, are built on the assumption that some underlying characteristic of a protein sequence, or of a protein structure, can be used to identify similar traits in related proteins. In other words, if part of a sequence or structure is preserved or conserved (whether because it is important to the activity of the protein, or because it is critical to its fold), this characteristic may be used to diagnose new family members. If such conserved traits are distilled from known protein families and stored in databases, then newly sequenced proteins may be rapidly analysed to determine whether they contain these previously recognised family characteristics. Searches of sequence pattern databases, and of fold template databases, are now routinely used to diagnose family relationships, and hence to infer structures and functions of newly determined sequences.

By definition, both sequence- and structure-based pattern recognition methods demand that a particular sequence or structure has been 'seen' before, and that some characteristic of it can be housed in a reference database. Sequence pattern recognition is far easier to achieve and is considerably more reliable than is fold recognition (which tends to give optimum results only in expert hands, and even then is still little better than 40% reliable). Nevertheless, both approaches are the subject of intensive research, and the methods continue to improve.

By contrast, prediction, the Holy Grail of bioinformatics, is still not possible, and is unlikely to be so for decades to come. Prediction stems from the idea that a functional site, or indeed a complete structure, need not have been 'seen' before, but can be deduced directly from the amino acid sequence, as illustrated in Figure 1.2. This approach obviates the need to create reference databases of functional site or structural templates, but requires instead the design of sophisticated software capable of meaningfully addressing the folding problem.

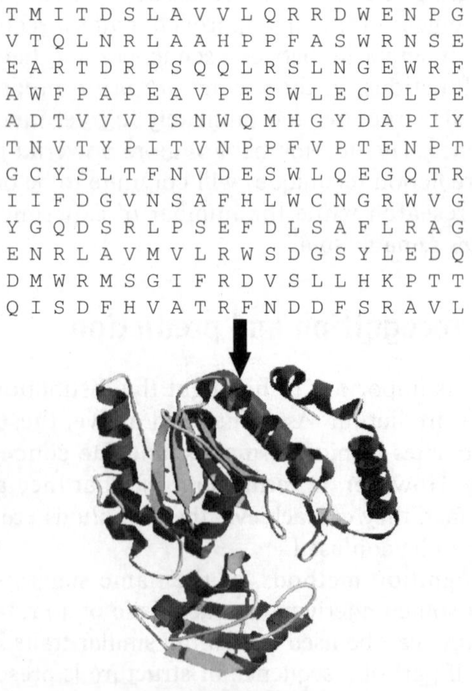

```
T M I T D S L A V V L Q R R D W E N P G
V T Q L N R L A A H P P F A S W R N S E
E A R T D R P S Q Q L R S L N G E W R F
A W F P A P E A V P E S W L E C D L P E
A D T V V V P S N W Q M H G Y D A P I Y
T N V T Y P I T V N P P F V P T E N P T
G C Y S L T F N V D E S W L Q E G Q T R
I I F D G V N S A F H L W C N G R W V G
Y G Q D S R L P S E F D L S A F L R A G
E N R L A V M V L R W S D G S Y L E D Q
D M W R M S G I F R D V S L L H K P T T
Q I S D F H V A T R F N D D F S R A V L
```

Figure 1.2 The Holy Grail of bioinformatics: the direct prediction of protein three-dimensional structure from the linear amino acid sequence.

1.9 The folding problem

The **folding problem** is a central theme of molecular biology. Simply stated, given the **primary structure** of a protein (see Box 1.2), how does the linear sequence of amino acids determine the final 3D fold? In 1961, Anfinsen showed that ribonuclease could be denatured and refolded without loss of enzymatic activity. This experiment suggested that all the information for a protein to adopt its native conformation is encoded in its primary structure. If this is really so, then it should theoretically be possible to derive the rules for protein folding from analyses of sequences with known structures, and hence to apply such rules to blind prediction of 3D structure, given only a linear sequence of amino acids.

At first sight, considering the ever-growing databases available to us, this might not seem to be an unrealistic expectation. However, in practice, in spite of more than three decades of research, the rules of protein folding have not been fully understood and structure prediction is still not possible (Rost and O'Donoghue, 1997). In 1998, methods of **secondary structure** prediction are little more than 50–60% reliable.

There are three main approaches to secondary structure prediction: (i) empirical statistical methods that use parameters derived from known 3D structures; (ii) methods based on physicochemical criteria (e.g., fold compactness, **hydrophobicity**, charge, hydrogen bonding potential, etc.); and (iii) prediction **algorithms** that use known structures of homologous proteins to assign secondary structure (Kyngas and Valjakka, 1998). One of the standard empirical statistical methods is that of Chou and Fasman, which is based on observed amino acid conformational preferences in non-homologous proteins. But in spite of being a 'standard' approach, like all other methods, its reliability is poor (~65%). This is because the sizes of the datasets used to derive the conformational potentials of the amino acids have been inadequate; and it is clear that they will remain inadequate for rigorous statistical analysis while the

---

BOX 1.2: LEVELS OF PROTEIN STRUCTURE

Primary structure:	the linear sequence of amino acids in a protein molecule
Secondary structure:	regions of local regularity within a protein fold (e.g., α-helices, β-turns, β-strands)
Super-secondary structure:	the arrangement of α-helices and/or β-strands into discrete folding units (e.g., β-barrels, $\beta\alpha\beta$-units, Greek keys, etc.)
Tertiary structure:	the overall fold of a protein sequence, formed by the packing of its secondary and/or super-secondary structure elements
Quaternary structure:	the arrangement of separate protein chains in a protein molecule with more than one subunit
Quinternary structure:	the arrangement of separate molecules, such as in protein–protein or protein–nucleic acid interactions

numbers of known structures of non-homologous proteins are so few. By contrast, for prediction algorithms, the use of multiple sequence data can improve matters, and may yield enhancements of several percent; but even a method that might, through judicious use of **sequence alignment** information, be claimed to be 70% reliable is of little practical value if, in a blind prediction, we do not know which 70% is correct.

Tertiary structure prediction (especially methods that build on secondary predictions, which, in practice, many do) is still further beyond reach. Indeed, as we learn more about the complexity of protein folding, the goal appears to recede further over the horizon. In 1998, it is clear that direct prediction of structure from sequence remains decades away.

1.10 The role of chaperones

A commonly used, but erroneous, argument against the idea that the linear amino acid sequence contains all the information necessary for protein folding is the presence of molecular **chaperones**. These are helper proteins that ensure that growing protein chains fold correctly (Hartl *et al.*, 1994). Chaperones are thought to block incorrect folding pathways that would lead to inactive products, by preventing incorrect aggregation and precipitation of unassembled **subunits**. They probably bind temporarily to interactive surfaces that are exposed only during the early stages of protein assembly. The point is, without their chaperones proteins still fold, but with them, many potential folding pathway cul-de-sacs are avoided, and hence they achieve their end-point with greater efficiency and fidelity.

1.11 Sequence analysis

The exact nature of the information encoded in the primary structure is unclear, and we still cannot read the language used to describe the final 3D fold of a biologically active macro-molecule. Indeed, detailed folding studies have revealed more and more complexities, making it clear that the sequence-to-structure relation is a hard problem (Gross, 1998). Nevertheless, there is a way forward. Using sequence analysis techniques, we can attempt to identify similarities between novel query sequences (i.e., whose structures and functions are unknown) and database sequences whose structures and functions have been elucidated. This is straightforward at high levels of sequence identity, where relationships are clear, but below 50% identity it becomes increasingly difficult to establish relationships reliably.

1.11.1 The Twilight Zone

In general, analyses can be pursued with decreasing certainty towards the **Twilight Zone** (Doolittle, 1986). This is a zone of sequence similarity (corresponding to ~0–20% identity) in which alignments may appear plausible to the eye, but are no longer statistically significant (in other words, the same alignment could have arisen by chance). To penetrate deeper into the Twilight Zone is the goal of most analytical methods. Many different approaches have been devised: some of these involve database searches with single sequences, others use characteristic chunks of sequence alignments; some weight database searches (using, say, **mutation** or **hydropathy** information), and others use only observed amino acid sequence data. Each method offers a slightly different perspective, depending on the type of information used in the search; none should be regarded as giving the right answer, or the full picture. The sensitivity ranges of some of these methods are depicted in Figure 1.3.

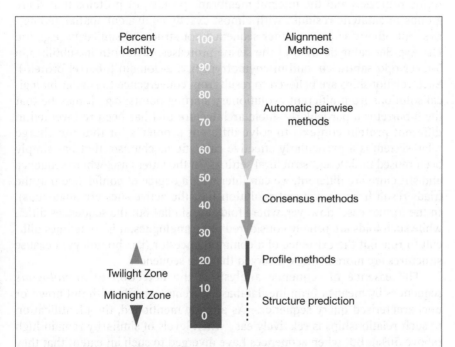

Figure 1.3 Application areas of different analysis methods. The scale indicates percent identity between two aligned sequences. Alignment of random sequences can produce around 20% identity; less than 20% does not constitute a significant alignment. Around this threshold is the Twilight zone, where alignments may appear plausible to the eye, but cannot be proved by current methods. Beyond the Twilight Zone is the so-called Midnight Zone, where sequence comparisons fail completely to detect structural similarities.

Until analysis methods are more reliable, it is important to use a range of techniques, slotting all the results together like pieces in a jigsaw. While the algorithms that recognise folds reliably, or that can predict structures, remain beyond our grasp, it is important to use the sequence analysis tools we have at our disposal, but in an intelligent way, always aware of their limitations. Chapter 8 discusses some of these methods in more detail.

1.12 Homology and analogy

Before moving on, it is important to define a concept that underpins the application of most sequence analysis methods, i.e., **homology**. The term homology, although easy to understand, is confounded and abused in the literature. Simply, sequences are said to be homologous if they are related by divergence from a common ancestor.

Understanding the meaning of homology allows us to appreciate the concept of analogy; this is encountered in the context of, say, protein structures that share similar folds but have no demonstrable sequence similarity (e.g., the ubiquitous β-barrel, found in such diverse proteins as the soluble serine proteases and the integral membrane porins); or proteins that share groups of catalytic residues with almost exactly equivalent spatial geometries, but otherwise have neither sequence nor structural similarity (e.g., the His-Asp-Ser catalytic triad of the serine proteases, seen both in subtilisin, a 3-layer αβα sandwich, and in chymotrypsin, a 2-domain β-barrel protein). Such relationships are believed to result from convergence to similar biological solutions from different evolutionary starting points: e.g., it may be that the β-barrel is a particularly stable architecture that has been re-invented in different protein contexts to solve different problems; or that the charge relay system is a particularly effective catalytic mechanism that has simply been reused in different structural settings. In the latter case, where sequence and structure are different, we can infer with a degree of confidence that the triads result from convergent evolution (i.e., the active sites are analogous). In the former case, however, where folds are similar but the sequences differ, while such folds are usually considered to be **analogues**, it is sometimes difficult to rule out the existence of a common ancestor (i.e., homology) because structures are more highly conserved than are sequences.

The essence of sequence analysis is the detection of *homologous* sequences by means of routine database searches, usually with unknown or uncharacterised query sequences. As already mentioned, the identification of such relationships is relatively easy when levels of similarity remain high (above 50%). But when sequences have diverged to such an extent that they are only 20% identical, or if two sequences share less than 20% identity, it becomes difficult or impossible to establish whether they might have arisen through the evolutionary processes of divergence or of convergence.

Homology is not a measure of similarity, but an absolute statement that sequences have a divergent rather than a convergent relationship. Thus, phrases that quantify homology (e.g., 'the sequences show 50% homology' or 'the sequences are highly homologous') are meaningless and should be avoided.

1.12.1 Orthology and paralogy

Among homologous sequences, it is useful to distinguish between proteins that perform the same function in different species (these are referred to as

orthologues) and those that perform different but related functions within one organism (so-called **paralogues**).

Sequence comparison of orthologous proteins opens the way to the study of molecular palaeontology. In particular cases, construction of **phylogenetic trees** has revealed relationships, for example, between proteins in bacteria, fungi and mammals, and between animals, insects and plants, inferences that are only unearthed by investigations at the molecular level. The study of paralogous proteins, on the other hand, has provided deeper insights into the underlying mechanisms of evolution. Paralogous proteins arose from single genes via successive duplication events. The duplicated genes have followed separate evolutionary pathways, and new specificities have evolved through variation and adaptation.

The mechanisms that resulted in the dispersal of paralogous proteins within genomes are diverse and often poorly understood. An example where multiple dispersal patterns are evident is the rhodopsin-like superfamily of G-protein-coupled receptors (Henikoff *et al.*, 1997). These proteins, which include light, olfactory, gustatory, hormone and neurotransmitter receptors, encompass a wide variety of functions. The emergence of different specificities and functions following gene duplication events may be detected by protein sequence comparison. For example, different visual receptors (opsins), which diverged from each other early in vertebrate evolution, are stimulated by different wavelengths of light. Human long-wavelength opsins (i.e., those sensitive to red and green light) are more closely related to each other (with around 95% sequence identity) than either sequence is to the short-wavelength blue-opsins, or to the rhodopsins (the achromatic receptors), with which they share an average 43% identity. The complexity that arises from the richness of such paralogous, and of orthologous, relationships presents a significant challenge for protein family classification.

1.13 The devil is in the detail

Much of the challenge of sequence analysis involves the marriage of biological information with sequence data. This process is made more difficult by the problem of orthology versus paralogy: following a database search, it is often unclear how much functional information can be legitimately transferred to the query sequence from a matched homologue. Many erroneous automatically derived functional annotations have now been incorporated into, and subsequently propagated through, sequence databases, because there is currently no quality assurance for functional annotation. In an era of information overload, where the use of computational tools is essential, this nevertheless highlights one of the dangers of reliance on fully automated procedures.

The analytical process is further complicated by the fact that, sometimes, sequence similarity is confined only to some part of an alignment. This scenario is encountered, in particular, when we study modular proteins. **Modules** may be thought of as a subset of protein **domains;** they are autonomous folding units that are contiguous in sequence, and are frequently used as protein building blocks. As building components, like Lego™ bricks, they may be used to confer a variety of different functions on the parent protein, either through multiple combinations of the same module, or via combinations of different modules to form **mosaics**. In genetic terms, the spread of modules cannot be explained simply by **gene duplication** and fusion events, but is thought to be the result of genetic shuffling mechanisms. Whatever the actual process, as elegantly described by Jacob in 1977, it appears that Nature behaves rather like a tinker, using a patchwork of existing components to produce a new, workable whole. Evolution, it seems, does not produce novelties from scratch, but works with old material, either transmogrifying a system to give it new functions, or combining several systems to produce a more elaborate one.

Reuse and integration of simple components are key to creating new, perhaps unexpected, functions in larger, more complex systems. The question then arises, is it possible to make predictions at a higher level, on the basis of what is known at a simpler one? The answer is that only limited conclusions may be drawn, because while the properties of a system can be explained by the properties of its components, they cannot be *deduced* from them. For the biologist, it may thus be impossible to predict, or even make an inspired guess at, the complex nature of a biochemical system from just a piecemeal understanding of some of its underlying molecular interactions.

In thinking about predictability, we can begin to get an appreciation of the difficulties ahead by considering a rather fantastic metaphor for biology encapsulated by Rube Goldberg machines (see Box 1.3). These are machines whose construction is so bizarre that an unknown 5% could not be deduced, even if 95% of the machine was known. Biology could be considered as the ultimate Goldberg machine. Jacob suggested that spare parts were grabbed during evolution's tinkering process and were misused (or adapted from their original functions), and then frozen by chance as integral pieces within a coherent and surviving organism. Traces, or echoes, of those earlier functions are preserved over time, but the remnants may be useless in surmising the new functions. Such ideas have important ramifications for today's would-be predictors, similarity searchers and detectors of evolutionary relationships. Clearly, the crux of the problem is in understanding the biological details; this is essential if we are to make sensible use of mathematics to infer relationships between components within complex systems, and if we are to have a hope of guessing at the bigger evolutionary picture.

In this book, we set out to describe some of the sequence comparison methods currently used to detect homologous (orthologous and paralogous) relationships, and examine some of the pitfalls. Notwithstanding the lessons of Goldberg machines, identifying evolutionary links between sequences *is* useful, as this often implies a shared function. At the level of the complete genome, diagnosis of protein function provides a vehicle by which the metabolic systems of different organisms can be compared, and as such is considered to be of more immediate value than is the prediction of protein

The American cartoonist Rube Goldberg invented implausible and fantastic machines. These machines contain historical, kinetic accidents, frozen in time, linked to other improbable components, glued together as complete systems, the functions of which appear to be miraculous. When only a small part of the machine is known (our current understanding of the machinery of biological systems), inductive completion of the entire mechanism is impossible.

In Goldberg's self-opening umbrella, the mechanism commences with rain falling on a dried prune; the moistened prune swells, activating a hand that flicks a lighter; this lights a candle that boils a kettle of water, the steam from which races through a whistle; the noise of the whistle frightens a monkey, who jumps onto a swing, the motion of which makes a blade cut a string that restrains a balloon; the freed balloon rises and, in so doing, opens the door to a bird cage; and this releases a host of tiny birds, each of which is tied to the ribs of a folded umbrella.

(RUBE GOLDBERG™ and © property of Rube Goldberg, Inc. Distributed by United Media)

Monkeys are exploited in many quite different Goldberg machines, where they perform different functions. In the self-opening umbrella, one could not predict a monkey sitting in exactly that spot, even with total knowledge of the rest of the components. To the extent to which biology functions like such a machine, protein homology searches are fraught with difficulties, especially when considering modular proteins, because: (a) revealing the presence of a module (monkey) tells us little or nothing of the function of the complete system; (b) knowing most of the components of a mosaic protein does not allow us to predict a missing module; and (c) modules found in different proteins do not necessarily perform precisely the same function.

structure. However, it must be remembered that between distantly related proteins, the 3D structures are more likely to be **conserved** than are the corresponding amino acid sequences. Indeed, many evolutionary relationships are apparent only at the level of shared structural features; such similarities cannot be detected even using the most sensitive sequence comparison methods (the region of identity where sequence comparisons fail completely to detect structural similarity has been termed the **Midnight Zone** (Rost, 1998)). Consequently, there is a theoretical limit to the effectiveness of sequence analysis techniques. Nevertheless, with this caveat, the following chapters set out to explain how to get the most from your protein sequence.

1.14 Summary

- The term bioinformatics is used to encompass almost all computer applications in biological sciences, but was originally coined in the mid-1980s for the analysis of biological sequence data.

- The quantity of known sequence data outweighs protein structural data by ~100:1 and, by virtue of the genome projects, sequence databases are doubling in size every year.

- A key challenge of bioinformatics is to analyse the wealth of sequence data in order to understand the amassed information in terms of protein structure, function and evolution.

- There are two principal analytical approaches in bioinformatics: i.e., pattern-recognition and prediction. Considerable progress has been made with pattern-recognition methods because of the availability of reference databases of sequence patterns and fold templates.

- Our incomplete understanding of the protein folding problem (i.e., how exactly the linear amino acid sequence determines the final 3D fold) presents a barrier to current attempts to predict structure directly from sequence.

- Homology is a central concept: sequences are said to be homologous if they are related by divergence from a common ancestor. Homology is not a synonym for similarity.

- The essence of sequence analysis is detection of homologous (orthologous (same function, different species) or paralogous (different but related functions within one organism)) relationships by means of routine database searches.

- The term analogy is used in the context of similar protein folds that share no detectable sequence similarity, or proteins that share groups of catalytic residues with the same spatial geometries, but otherwise have no sequence or structural similarity. Such relationships are thought to result from the evolutionary process of convergence.

- Searches can be pursued with decreasing certainty towards the Twilight Zone, where alignments are no longer statistically significant. In the

Midnight Zone, sequence comparisons fail completely to detect structural similarities.

17

Further reading

- Wherever possible, a range of different analysis methods should be used, and the results should be married with all available biological information.

1.15 Further reading

Protein sequencing
EDMAN, P. (1950) *Acta Chem. Scand.*, **4**, 283–293.

EDMAN, P. and BEGG, G. (1967) A protein sequenator. *European Journal of Biochemistry*, **1**, 80–91.

HIRS, C.H.W., MOORE, S. and STEIN, W.H. (1960) *Journal of Biological Chemistry*, **235**, 633–647.

RYLE, A.P., SANGER, F., SMITH, L.F. and KITAI, R. (1955) *Biochemical Journal*, **60**, 541–556.

Sequence analysis
DOOLITTLE, R.F. (1986) *Of URFs and ORFs: A Primer on How to Analyse Derived Amino Acid Sequences.* University Science Books, Mill Valley, CA.

HENIKOFF, S., GREENE, E.A., PIETROKOVSKI, S., BORK, P., ATTWOOD, T.K. and HOOD, L. (1997) Gene families: The taxonomy of protein paralogs and chimeras. *Science*, **278**, 609–614.

Evolution
GOLD, L., SINGER, B., HE, Y.-Y. and BRODY, E. (1997) SELEX and the evolution of genomes. *Current Opinion in Genetics and Development*, **7**, 848–851.

JACOB, F. (1977) Evolution and tinkering. *Science*, **196**, 1161–1166.

Protein folding and structure prediction
ANFINSEN, C.B. *et al.* (1961) *Proceedings of the National Academy of Science*, **47**, 1308–1314.

GROSS, M. (1998) Protein folding: Think globally, (inter)act locally. *Current Biology*, **8**, R308–R309.

HARTL, F.-U., HLODAN, R. and LANGER, T. (1994) Molecular chaperones in protein folding: the art of avoiding sticky situations. *TiBS*, **19**, 20–25.

KYNGAS, J. and VALJAKKA, J. (1998) Unreliability of the Chou–Fasman parameters in predicting protein secondary structure. *Protein Engineering*, **11**, 345–348.

ROST, B. (1998) Marrying structure and genomics. *Structure*, **6**, 259–263.

ROST, B. and O'DONOGHUE, S. (1997) Sisyphus and prediction of protein structure. *Computer Applications in the Biosciences*, **13**(4), 345–356.

RUSSELL, R.B. and PONTING, C. (1998) Protein fold irregularities that hinder sequence analysis. *Current Opinion in Structural Biology*, **8**, 364–371.

Databases

BERNSTEIN, F.C., KOETZLE, T.F., WILLIAMS, G.J.B., MEYER, E.F., BRICE, M.D., RODGERS, J.R., KENNARD, O., SHIMONOUCHI, T. and TASUMI, M. (1977) The Protein Data Bank: A computer based archival file for macromolecular structures. *Journal of Molecular Biology*, **112**, 535–542.

BOGUSKI, M.S., TOLSTOSHEV, C.M. and BASSETT, D.E. (1994) Gene discovery in DBEST. *Science*, **265**, 1993–1994.

Genomes

BULT, C.J., WHITE, O., OLSEN, G.J., ZHOU, L.X., FLEISCHMANN, R.D., SUTTON, G.G., BLAKE, J.A., FITZGERALD, L.M., CLAYTON, R.A., GOCAYNE, J.D., KERLAVAGE, A.R., DOUGHERTY, B.A., TOMB, J.F., ADAMS, M.D., REICH, C.I., OVERBEEK, R., KIRKNESS, E.F., WEINSTOCK, K.G., MERRICK, J.M., GLODEK, A., SCOTT, J.L., GEOGHAGEN, N.S.M., WEIDMAN, J.F., FUHRMANN, J.L., NGUYEN, D., UTTERBACK, T.R., KELLEY, J.M., PETERSON, J.D., SADOW, P.W., HANNA, M.C., COTTON, M.D., ROBERTS, K.M., HURST, M.A., KAINE, B.P., BORODOVSKY, M., KLENK, H.P., FRASER, C.M., SMITH, H.O., WOESE, C.R. and VENTER, J. (1996) Complete genome sequence of the Methanogenic Archaeon, *Methanococcus jannaschii*. *Science*, **273**, 1058–1073.

FLEISCHMANN, R.D., ADAMS, M.D., WHITE, O., CLAYTON, R.A., KIRKNESS, E.F., KERLAVAGE, A.R., BULT, C.J., TOMB, J.F., DOUGHERTY, B.A., MERRICK, J.M., MCKENNEY, K., SUTTON, G., FITZHUGH, W., FIELDS, C., GOCAYNE, J.D., SCOTT, J., SHIRLEY, R., LIU, L.I., GLODEK, A., KELLEY, J.M., WEIDMAN, J.F., PHILLIPS, C.A., SPRIGGS, T., HEDBLOM, E., COTTON, M.D., UTTERBACK, T.R., HANNA, M.C., NGUYEN, D.T., SAUDEK, D.M., BRANDON, R.C., FINE, L.D., FRITCHMAN, J.L., FUHRMANN, J.L., GEOGHAGEN, N.S.M., GNEHM, C.L., MCDONALD, L.A., SMALL, K.V., FRASER, C.M., SMITH, H.O. and VENTER, J.C. (1995) Whole-genome random sequencing and assembly of *Haemophilus influenzae* Rd. *Science*, **269**, 496–512.

FRASER, C.M., GOCAYNE, J.D., WHITE, O., ADAMS, M.D., CLAYTON, R.A., FLEISCHMANN, R.D., BULT, C.J., KERLAVAGE, A.R., SUTTON, G., KELLEY, J.M., FRITCHMAN, J.L., WEIDMAN, J.F., SMALL, K.V., SANDUSKY, M., FUHRMANN, J., NGUYEN, D., UTTERBACK, T.R., SAUDEK, D.M., PHILLIPS, C.A., MERRICK, J.M., TOMB, J.F., DOUGHERTY, B.A., BOTT, K.F., HU, P.C., LUCIER, T.S., PETERSON, S.N., SMITH, H.O., HUTCHISON, C.A. and VENTER, J.C. (1995). The minimal gene complement of *Mycoplasma genitalium*. *Science*, **270**, 397–403.

GOFFEAU, A., BARRELL, B.G., BUSSEY, H., DAVIS, R.W., DUJON, B., FELDMANN, H., GALIBERT, F., HOHEISEL, J.D., JACQ, C., JOHNSTON, M., LOUIS, E.J., MEWES, H.W., MURAKAMI, Y., PHILIPPSEN, P., TETTELIN, H. and OLIVER, S.G. (1996) Life with 6000 genes. *Science*, **274**, 546–567.

KOONIN, E.V., TATUSOV, R.L. and GALPERIN, M.Y. (1998) Beyond complete genomes: from sequence to structure and function. *Current Opinion in Structural Biology*, **8**, 355–363.

CHAPTER TWO

Information networks

2.1 Introduction

The purpose of this chapter is to introduce the Internet, the World Wide Web, and the global network of biological information and service providers, the interplay between which has made the bioinformatics revolution possible. The chapter refers briefly to the advent of the browsers that have made such information and facilities accessible, and to the software tools that have been developed for linking and searching distributed databases. The Web addresses of some important bioinformatics centres are summarised.

2.2 What is the Internet?

The **Internet** is a global network of computer networks that links government, academic and business institutions. In order to work effectively, the networks share a **communication protocol**, called **Transmission Control Protocol/Internet Protocol**, better known as **TCP/IP**. Such a shared mechanism of communication means that different types of machine are able to speak to each other in a common way.

Computers within the network are referred to as nodes, and these communicate with each other by transferring data **packets**. But packets do not necessarily travel directly from one machine to another; they may pass through several computers en route to their final destination. If any of the nodes on the way are **down**, the network protocols are designed to find an alternative route.

2.3 How do computers find each other?

To facilitate communication between nodes, each computer on the Internet is given a unique, identifying number (its **IP address**), which is encoded in a dotted-decimal format. So, for example, one node on the Internet might have the IP address:

128.40.46.17

These numbers were designed to be intelligible to computers, and, as a result, are not at all people-friendly. But an alternative, hierarchical domain-name system has also been implemented, which makes Internet addresses easier to decipher. The name identifies, in turn, the particular machine, the site where the machine lives, and the domain (and subdomain) to which the site belongs (some of the various Internet domains and subdomains are listed in Table 2.1). For example, the above numerical address translates as:

bsmir30.biochemistry.ucl.ac.uk

This tells us that the machine is called bsmir30, which lives in the Biochemistry Department at University College London (UCL), which belongs to the academic (ac) subdomain of the UK country domain.

Table 2.1 Example Internet domains and subdomains

Country-based domains		Other domains		Subdomains	
Australia	.au	Educational	.edu	Academic	.ac
Denmark	.dk	Commercial	.com	Company	.co
Finland	.fi	Governmental	.gov	Other organisation	.org
France	.fr	Military	.mil	General	.gen
Germany	.de				
Greece	.gr				
Hungary	.hu				
Ireland	.ie				
Israel	.il				
Italy	.it				
Netherlands	.nl				
New Zealand	.nz				
Poland	.pl				
Portugal	.pt				
South Africa	.za				
Spain	.es				
Sweden	.se				
Switzerland	.ch				
United Kingdom	.uk				
USA	.us				

2.4 Facilities used on the Internet

It is interesting to discover what people most commonly use the Internet for. The most familiar services include electronic mail (email), newsgroups, file transfer and remote computing. The first of these concern communication between people, whether on a one-to-one basis, as with email, or at the level of group discussions, via newsgroups. The other facilities are largely concerned with remote computing, involving the use, for example, of the **File Transfer Protocol (FTP)** to transfer files between machines, and the **Telnet protocol**, by which users may connect to computers at different sites and use the machines as if physically present at the remote location.

Amongst the most exciting of Internet services are those that permit communication between users in real-time. These include the UNIX talk protocol (or VMS phone), which is analogous to holding a telephone conversation, but users 'speak' to each other by typing into a shared screen. An extension of this concept is conferencing, whereby *groups* of people meet and 'talk' to each other, again by typing into a shared interface (e.g., such as provided by the WebBoard facility and/or the BioMoo MultiUser Dungeon); this is reckoned to be the virtual analogue of meeting colleagues for scientific discussions over coffee! Notwithstanding the excitement of such virtual innovations, however, they can be cumbersome, confusing and slow to use, and, like virtual coffee, are not quite as good as the real thing. Thus, email remains one of the most popular of the Internet services.

2.5 What is the World Wide Web?

The **World Wide Web** (the Web, WWW or W3) was conceived and developed at CERN, the European Laboratory for Particle Physics, to allow information sharing between internationally dispersed groups in the High Energy Physics community. The concept of information sharing between remote locations, and the ramifications for rapid data dissemination and communication, found immediate applications in numerous other areas. As a result, the Web spread quickly and is now making a profound impact in the field of bioinformatics. Today, the WWW is the most advanced information system deployed on the Internet.

The Web is a hypermedia-based information system. So popular and powerful has it become that it is now almost synonymous with the Internet itself. On the W3 Consortium home page, the Web is described as 'the universe of network-accessible information, the embodiment of human knowledge' (a slightly exaggerated claim perhaps, but one that may well be met in the new millennium).

2.6 Web browsers

The full potential of the Internet was only properly realised with the advent of **browsers**, which for the first time allowed easy access to information at different sites. Browsers are **clients** that communicate with **servers**, using a set of standard protocols and conventions. The first point of contact between a browser and a server is the **home page.** The default home pages invoked by particular browsers tend to point to the software companies of their respective manufacturers, but they may be easily customised to point to more useful, frequently visited sites, or to the user's own home page.

Once the browser has loaded its initial page, it then provides an easy-to-use interface with which to retrieve documents, access files, search databases, and so on. Some of the most commonly used browsers are outlined below.

2.6.1 Lynx

Lynx was developed in the Academic Computing Services at the University of Kansas, USA, as part of an effort to build a campus-wide information system. It runs on UNIX or VMS **operating systems**, providing a text-only interface via low-cost, **dumb** display devices, such as the ubiquitous VT100 terminal (or emulator). It is probably the most widely used text-only browser on the Internet. Although it may seem strange that a text-mode browser could be preferred over a graphical interface (particularly in the era of the multimedia environment), in some circumstances Lynx is more efficient than graphical browsers, which use large amounts of computer resources (e.g., where a comparatively slow dial-up connection is in use).

2.6.2 Mosaic

Mosaic was developed in 1993 at the National Center for Supercomputing Applications (NCSA), University of Illinois, Urbana-Champaign, USA. As a hypermedia system designed for X-Windows, Apple Mac and Microsoft Windows platforms, it provided a single, user-friendly interface to the diverse protocols, data formats and information servers available throughout the Internet. At an early stage, Mosaic was undoubtedly responsible for the surge in popularity of the Web, and it was hard to imagine that such a 'killer application, the utility that will bring the Internet to the masses' could be dethroned. But the WWW is in a constant state of change, and the tools that interact with it are developing at an incredible rate. As it turned out, Mosaic's monopoly was short-lived.

2.6.3 Netscape Navigator

Netscape Navigator was developed in 1994 by Netscape Communications Corporation, Mountain View, California, USA. It was designed as an alternative to Mosaic, and was almost an overnight success. It is now the most popular package for browsing information on the Internet – it has been estimated that more than 80% of Internet users browse the Web with Netscape. Current versions of the software include facilities such as Internet email, frames, real-time communication, audio and video support, and the latest technology to support creation of visually exciting, fully interactive pages (e.g., with **Java** applets). An example Web page displayed within the Netscape browser is shown in Figure 2.1.

2.6.4 Internet Explorer

Internet Explorer was developed in 1995 by Microsoft Corporation, Redmond, USA. It was based on NCSA Mosaic and is designed to work

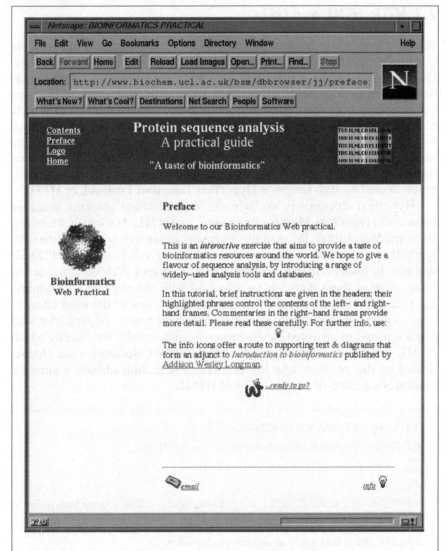

Figure 2.1 Example Web page displayed from the Netscape browser, illustrating the use of frames to manage different aspects of a document in different windows. Also illustrated is the use of hyperlinks to other documents or images via highlighted words and phrases, the centring and customisation of the size and colour of text, the embedding of images, and so on. The featured page is the home page of the interactive bioinformatics practical that accompanies this text.

with PC-based operating systems. It offers the familiar functionality of other hypermedia browsers, including support for frames, Java and ActiveX. Originally developed as a Windows95/NT product, current versions also run on Sun's version of UNIX.

2.7 HTTP, HTML and URLs

The documents that browsers display exploit **hypertext** and **hypermedia** techniques to make Web browsing and publishing extremely easy. Hypertext documents contain text with embedded links (so-called **hyperlinks**) to other documents. Hyperlinks are usually characterised by being highlighted in some way, either using a different colour from the main body of the text (see Figure 2.1), or by being boxed, etc.. Selecting a highlighted link calls up the linked document, regardless of its location, whether on the same server, or on a server in a different country. Communication between hyperlinks is transparent; the name given to the underlying communication protocol used by Web servers is **HyperText Transport Protocol**, or **HTTP.**

Hypertext documents are written in a standard markup language known as **HyperText Markup Language**, or **HTML**. Markup instructions allow the Web author, for example, to render phrases in bold type (the <B> symbol), to customise the size and colour of the font (the <FONT> symbol), to insert horizontal rulers (<HR>), images (<IMG>), and so on (note: each of these modes is switched off with the relevant </> symbol, e.g. </B>). HTML is simple and quick to learn – one of the most effective ways to gain experience is to view the 'document source' of particular Web pages (an option provided by the browser), and thereby see directly which HTML commands result in specific effects. HTML documents are characterised by the .html or .htm file extensions (e.g., index.html). Figure 2.2 illustrates a section of a typical page of HTML.

```
<TITLE> BIOINFORMATICS PRACTICAL </TITLE>
<CENTER><H2>"A taste of bioinformatics"</H2></CENTER>
<P>
<HR>
<P>
<FONT COLOR="#FF0000" SIZE=+1><B>Welcome</B></FONT> to our Web practi-
cal on bioinformatics, a gentle introduction to global sequence and structure analysis
facilities. We hope the experience will reveal the power and potential both of W3 and
of bioinformatics, but most of all we hope you have fun!
<P>

Highlighted phrases provide access either to additional pages within the tutorial, or to
"helpful" pictures that explain the text in more detail. Links to pictures are denoted by
a green blob <IMG SRC="http://www.biochem.ucl.ac.uk/bsm/ dbbrowser/c32/ablob.
gif"> – occasional very important pictures have a red blob <IMG SRC="http://www.
biochem.ucl.ac.uk/bsm/dbbrowser/c32/cblob.gif">.
<P>
<HR>
<P>
<A HREF=mailto:attwood@biochemistry.ucl.ac.uk><I>attwood@biochemistry.ucl.ac.
uk</I></A>
```

Figure 2.2 Excerpt from an HTML file, showing typical mark-up instructions, including text centring, image insertion, etc. The extract is taken from the HTML document that produced the original home page of the Web practical.

Documents are accessed, like computers themselves, by means of unique addresses, or **Uniform Resource Locators (URLs)**. The URL contains a number of distinct parts, which identify, in turn, the underlying communication protocol, the Web server, often a directory path, and sometimes a file name. For example:

http://www.biochem.ucl.ac.uk/bsm/dbbrowser/jj/prefacefrm.html

This URL identifies the communication protocol as HTTP, it points to the Web server in the Biochemistry Department at University College London, and gives the directory path that points to the hypertext document prefacefrm.html.

2.8 The European Molecular Biology network – EMBnet

Long before browsers made accessible, and consequently popularised, the Internet, organisations around the world saw the potential of the Internet as a force for global communication and resource centralisation. This became particularly important in the mid-1980s as biological databases were beginning to proliferate and users were demanding more efficient means of access to more up-to-date data.

In 1988, a network was established to link European laboratories that used biocomputing and bioinformatics in molecular biology

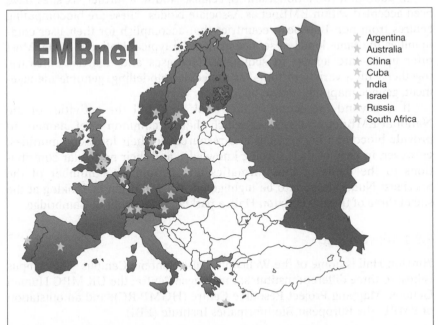

Figure 2.3 The European Molecular Biology Network (EMBnet) of bioinformatics and genomics centres. In 1998, EMBnet operates 20 National, eight Specialist and six Associate Nodes.

research. The network, known as EMBnet, was envisaged as a way of providing information, services and training to users in dispersed European laboratories, via designated nodes operating in their local languages – Figure 2.3. The establishment of such centralised national facilities was an important step: as the field of computational biology expanded, it removed the necessity for individual institutions to keep up-to-date copies of a range of biological databases (and to buy the disk space required to accommodate them), to install the corresponding range of search tools, and/or to buy expensive licences to use commercial software packages to access the data.

In 1998, 10 years after its inception, EMBnet operates 34 nodes, as detailed in Table 2.2. Of these, 20 are designated National Nodes. These are appointed by the governments of their respective nations, and have a mandate to provide databases, software and on-line services (including sequence analysis, protein modelling, genetic mapping, etc.); to offer user support and training; and to undertake research and development (such as the design of the Sequence Retrieval System (SRS)).

A further eight EMBnet nodes are 'Specialist' sites. These are academic, industrial or research centres that are considered to have particular knowledge of specific areas of bioinformatics. They are largely responsible for the maintenance of biological databases and software: e.g., the EBI maintains the EMBL nucleotide database, the ICGEB maintains the SBASE annotated domain database, and so on.

In adddition to National and Specialist Nodes, a further six sites have been accepted within EMBnet as Associate Nodes. These are biocomputing centres from non-European countries that accomplish for their user communities the same kinds of service as might a typical National Node. Most offer up-to-date access to sequence databases and analysis software, together with a variety of tools for molecular modelling, genome management, genetic mapping, and so on.

It is beyond the scope of this book to detail the activities of the National EMBnet nodes, all of which share a common brief, namely, to provide biocomputing services and resources to their local communities. However, in view of their specific knowledge and their particular contributions to the fields of bioinformatics and genomics, a number of the Specialist Nodes deserve to be highlighted. We will begin by looking at the site of three of these – Hinxton Hall, a parkland site south of Cambridge.

2.8.1 Hinxton Hall

Hinxton Hall is home of the Wellcome Trust Genome Campus. The campus is host to three different institutes: the Sanger Centre; the UK MRC Human Genome Mapping Project Resource Centre (HGMP-RC); and an outstation of EMBL, the European Bioinformatics Institute (EBI).

The Sanger Centre
The Sanger Centre is a genome research centre, established in 1992 by the Wellcome Trust and the Medical Research Council. The overall role of the

EMBnet National Nodes

Vienna Biocenter	Austria	http://www.at.embnet.org/
BEN	Belgium	http://www.be.embnet.org/
BioBase	Denmark	http://biobase.dk/
CSC	Finland	http://www.fi.embnet.org/
INFOBIOGEN	France	http://www.infobiogen.fr/
GENIUSnet	Germany	http://genome.dkfz-heidelberg.de/biounit/
IMBB	Greece	http://www.imbb.forth.gr/
HEN	Hungary	http://www.hu.embnet.org/
INCBI	Ireland	http://acer.gen.tcd.ie/
INN	Israel	http://dapsas.weizmann.ac.il/bcd/inn.html
IEN-ADR	Italy	http://bio-www.ba.cnr.it:8000/BioWWW/Bio-WWW.htm
CAOS/CAMM	Netherlands	http://www.caos.kun.nl/
Bio	Norway	http://www.no.embnet.org/
IBB	Poland	http://www.ibb.waw.pl/
PEN	Portugal	http://www.pen.gulbenkian.pt/
GeneBee	Russia	http://www.genebee.msu.su/
CNB-CSIC	Spain	http://www.es.embnet.org/
BMC	Sweden	http://www.embnet.se/
SIB	Switzerland	http://www.ch.embnet.org/
SEQNET	UK	http://www.seqnet.dl.ac.uk/

EMBnet Specialist Nodes

MIPS	Germany	http://www.mips.biochem.mpg.de/
ICGEB	Italy	http://www.icgeb.trieste.it/
Pharmacia Upjohn	Sweden	http://www.pnu.com/
F.Hoffmann–La Roche	Switzerland	http://www.roche.com/
EBI	UK	http://www.ebi.ac.uk/
HGMP-RC	UK	http://www.hgmp.mrc.ac.uk/
Sanger	UK	http://www.sanger.ac.uk/
UCL	UK	http://www.biochem.ucl.ac.uk/bsm/dbbrowser/embnet/

EMBnet Associate Nodes

IBBM	Argentina	http://sol.biol.unlp.edu.ar/
ANGIS	Australia	http://www.angis.su.oz.au/
CBI	China	http://www.cbi.pku.edu.cn/
CIGB	Cuba	http://bio.cigb.edu.cu/
CDFD	India	Not yet available
SANBI	South Africa	http://www.sanbi.ac.za

USA Information Providers

NCBI	USA	http://www.ncbi.nlm.nih.gov/
NLM	USA	http://www.nlm.nih.gov/
NIH	USA	http://www.nih.gov/

Centre is to further our knowledge of genomes, focusing especially on mapping and sequencing the human genome. By 2002, the Centre hopes to have obtained one sixth of the 3000 million base human genome sequence, at a cost of £50 million. Completion of the sequence is expected to be achieved by ~2005 via an international collaboration of sequencing centres.

As a pilot for the human genome project, the Centre has been involved in a collaboration with the Genome Sequencing Center at Washington University, St Louis, USA, aiming to sequence the complete genome of the model organism *Caenorhabditis elegans* (this is 1/30 the size of the human genome). The Centre is also sequencing DNA from various microorganisms, including *Saccharomyces cerevisiae*, *Schizosaccharomyces pombe* and human pathogens such as the tuberculosis bacterium.

The UK MRC Human Genome Mapping Project – Resource Centre

The HGMP-RC is funded by the UK Medical Research Council. It is active in the Human and Mouse Genome Projects, playing a role as both a materials and a service provider. It also provides an online computing service, offering user support and training courses. The specific brief of the Centre is to provide data and services to the medical research community, and to facilitate genomic research by providing centralised training resources. In 1999, the UK National node, SEQNET, which is currently hosted by the Daresbury Laboratory in Warrington, will relocate to this site.

The European Bioinformatics Institute – EBI

Established in 1994, the European Bioinformatics Institute (EBI) (Emmert *et al.*, 1994) is an outstation of the European Molecular Biology Laboratory (EMBL), an international research institute with its headquarters in Heidelberg, Germany. One of the central activities of the EBI is development and distribution of the EMBL Nucleotide Sequence Database, Europe's primary nucleotide sequence data resource. This is a collaborative project with GenBank (NCBI, Bethesda, USA) and DDBJ, the DNA Database of Japan (Mishima, Japan), which ensures that all new and updated database entries are shared between the groups on a daily basis. The EBI also collaborates with the Swiss Institute of Bioinformatics to maintain and distribute the SWISS-PROT protein sequence database. More than 30 additional specialist molecular biology databases are also distributed through EBI releases and network services.

2.8.2 MIPS

The Martinsried Institute for Protein Sequences (MIPS) (Max-Planck Institut für Biochemie, Germany) is the European partner of the PIR-International Protein Sequence Database. The role of MIPS is to collect, distribute and maintain up-to-date protein sequence data within Europe. The Institute also provides access to a database of aligned protein families, and to dynamic database search software for retrieval of homologues and inspection of alignments. MIPS also plays an active role in the field of Genome Research in its guise as informatics co-ordinator for the *Saccharomyces cerevisiae* and *Arabidopsis thaliana* Projects of the European Commission.

2.8.3 UCL

The Biomolecular Structure and Modelling (BSM) unit at University College London is a biocomputing centre with expertise in two central areas of bioinformatics, specifically in protein sequence and structure analysis. In 1998, the unit was host to groups responsible for the maintenance of the PRINTS protein fingerprint database; the PDBsum database, which provides summaries and structural analyses of PDB data files; and the CATH protein structure classification resource. The unit also makes a variety of analysis tools available from its FTP site, including programs to plot schematic diagrams of protein–ligand interactions; to calculate hydrogen- and non-bonded interactions; to analyse protein structural motifs; and to check the stereochemical quality of protein structures. Other programs are available for direct use via the Web, including a variety of database search and sequence analysis tools, and interactive sequence alignment and structure visualisation software. These facilities are provided via the DbBrowser Bioinformatics Web server (Michie *et al.*, 1996), which, by 1999, is expected to have relocated to the University of Manchester.

2.8.4 The Sequence Retrieval System – SRS

Although EMBnet has played a vital role in centralising data resources for its national user communities, a problem that emerged was that there was no effective, efficient way of interrogating all the resources gathered together at a particular site, since there were no common formats among the different database types. As a result, a research project was undertaken within EMBnet to address the problems inherent in interfacing complex environments. The resulting product was the Sequence Retrieval System, SRS, a network browser for databases in molecular biology (Etzold and Argos, 1993). SRS allows any flat-file database to be indexed to any other. This has the advantage that the derived indices may be rapidly searched, allowing users to retrieve, link and access entries from all the interconnected resources. The system has the particular strength that it can be readily customised to use any defined set of databanks. Typically, the resource links nucleic acid, EST, protein sequence, protein pattern, protein structure, specialist/boutique and/or bibliographic databases, as shown in Figure 2.4. SRS is thus a very powerful tool, allowing users to formulate queries across a range of different database types via a single interface, without having to worry about underlying data structures, query languages, and so on.

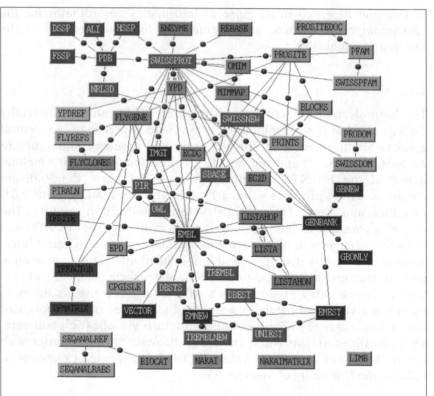

Figure 2.4 Typical network of databases linked via the SRS Sequence Retrieval System,
illustrating the potential complexity of queries that may be built up via the labyrinth of
relationships between biological and bibliographic databases. The ability to link such diverse
data types renders SRS an extremely powerful database interrogation tool.

2.9 The National Center for Biotechnology Information – NCBI

Having discussed the various European initiatives to maintain biological
data repositories and provide biocomputing services, it is appropriate to
mention the leading American information provider, the National Center
for Biotechnology Information (NCBI). The NCBI was established in 1988
as a division of the National Library of Medicine (NLM), and is located on
the campus of the National Institutes of Health (NIH) in Bethesda,
Maryland. The NLM was chosen to host the NCBI because of its experience in biomedical database maintenance, and because, as part of the NIH,
it could establish a research programme in computational biology.

The role of the NCBI is to develop new information technologies to aid
our understanding of the molecular and genetic processes that underlie
health and disease. Its specific aims include the creation of automated systems for storing and analysing biological information; the development of
advanced methods of computer-based information processing; the facilita-

tion of user access to databases and software; and the co-ordination of efforts to gather biotechnology information worldwide.

Since 1992, one of the principal tasks of the NCBI has been the maintenance of GenBank, the NIH DNA sequence database. Groups of annotators create sequence data records from the scientific literature and, together with information acquired directly from authors, data are exchanged with the international nucleotide databases, EMBL and DDBJ.

2.9.1 Entrez

Just as in Europe the SRS system was designed to facilitate access to a range of bio-databanks, so at the NCBI the Entrez facility was developed to allow retrieval of molecular biology data and bibliographic citations from NCBI's integrated databases (Schuler *et al.*, 1996). Entrez uses a rather different and, in some ways, slightly less flexible approach from SRS that does not allow customisation with an institution's preferred databases. Nevertheless, perhaps its most valuable feature is its exploitation of the concept of 'neighbouring', which allows related articles in different databases to be linked to each other, whether or not they are cross-referenced directly. Entrez typically provides access to DNA sequences (from GenBank, EMBL and DDBJ); protein sequences (from SWISS-PROT, PIR, PRF SEQDB, PDB and translated protein sequences from the DNA sequence databases); genome and chromosome mapping data; 3D protein structures from PDB; and the PubMed bibliographic database.

The development of facilities such as SRS and Entrez to integrate and access multiple databases is important: first, such systems include different resources and hence allow different perspectives on the same, or similar, queries; and second, the 'traffic' problems that have resulted from the popularity of the Web often mean that remote servers are almost impossible to contact – at such times, to have different trans-Atlantic alternatives for database interrogation is highly advantageous.

2.10 Virtual tourism

The very success of the WWW has, in recent years, posed serious problems for those wishing to use it 'seriously'. The Web is not only bearing the burden of work-related traffic, but is now also carrying a more leisure-related cargo. In short, the Information Superhighway is becoming clogged with the caravans and RVs of the Virtual Tourist.

The consequences of virtual tourism are far-reaching. As the name suggests, facilities are now available that permit information retrieval from anywhere in the world. There is no end to the type of information available: from the spheres of business, current affairs, education, entertainment, finance, shopping, sport, travel, etc. Navigation to different corners of the virtual globe may be effected by means of a variety of search engines, clickable maps, and so on.

One such facility is provided by The Virtual Tourist, which offers a geographic directory of WWW servers in the world. Here, information is presented in the form of clickable world, country and state maps. For example, to locate a particular biocomputing centre in Australia, the would-be tourist simply clicks on the continent of Australia in the world map, which leads to the Australia and Pacific map. Clicking on the continent yields an interactive map of sites by state. Choosing New South Wales provides the map for this territory, and clicking on the city of Sydney then offers a directory of all Web servers in Sydney. Within this directory, the tourist finds a list of universities, and under the heading of the University of Sydney is found the Australian National Genomic Information Service (ANGIS) – the Australian node of EMBnet. Travel was never easier! Snapshots of this Australian tour are shown in Figure 2.5.

For information about countries, states and regions, an alternative to the Virtual Tourist is offered by City.Net which provides global tourist information.

Facilities like these make the Web so appealing that organisations and individuals alike have embraced the Information Age and the Internet with a

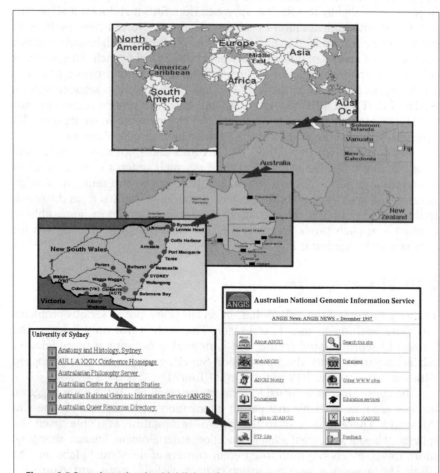

Figure 2.5 Snapshots showing highlights of a virtual trip to Australia, navigating by means of clickable world, continent and state maps, and directories of Web servers, until reaching the final destination, ANGIS, the Australian National Genomic Information Service.

vengeance. But two specific problems arise from this popularity: (i) there is a growing frustration for people in the workplace when their Internet excursions meet 'rush-hour' traffic, as their trans-Atlantic counterparts wake up and the Web literally grinds to a halt; and (ii) a more sinister aspect is that 'anything goes' on the Internet, including pornography, which poses genuine problems for schools that wish to make use of Internet facilities, without risk to students.

In the late 1990s, we are still in a honeymoon period, when use of the Internet is not regulated. But it is likely that this situation will not last forever, and we may eventually see regulation at some level. As private use of the Internet booms, some form of central control may have to be established, both to get government funds invested in increased bandwidth, and to establish a formal backbone structure for which the user, finally, must pay.

2.11 Summary

- The Internet is a global network of computer networks. Each computer, or network node, has a unique address by which it can be identified and can communicate with other nodes.

- The Internet provides many services, the most popular of which include email, newsgroups, file transfer and remote computing.

- The World Wide Web (WWW) is the most advanced information system on the Internet, and is so powerful that it has become almost synonymous with the Internet itself.

- Browsers provide easy-to-use interfaces for accessing information on the Web. The first point of contact between a browser and a Web server is the home page.

- Documents that browsers display are accessed by means of unique addresses, so-called Uniform Resource Locators, or URLs.

- EMBnet is a network of European biocomputing laboratories. Its National Nodes provide on-line services, user support and training; its Specialist Nodes provide databases and software. Three key Specialist Nodes are the Sanger Centre, the HGMP-RC, and the EBI, home of the EMBL, SWISS-PROT and TrEMBL databases.

- The SRS Sequence Retrieval System was developed within EMBnet to allow information retrieval across a range of different database types via a single interface.

- The leading American bio-information provider is the NCBI, home of the GenBank database and the Entrez information retrieval system.

2.12 Further reading

The Internet
SWINDELL, S.R., MILLER, R.R. and MYERS, G.S.A. (eds) (1996) *Internet for the Molecular Biologist.* Horizon Scientific Press.

Information retrieval tools
ETZOLD, T. and ARGOS, P. (1993) SRS an indexing and retrieval tool for flat file data libraries. *Computer Applications in the Biosciences*, **9**, 49–57.

SCHULER, G.D., EPSTEIN, J.A., OHKAWA, H. and KANS, J.A. (1996) Entrez: molecular biology database and retrieval system. *Methods in Enzymology*, **266**, 141–162.

Bioinformatics WWW servers
APPEL, R.D., BAIROCH, A. and HOCHSTRASSER, D.F. (1994) A new generation of information-retrieval tools for biologists – the example of the ExPASy WWW server. *TiBS*, **19**(6), 258-260.

MICHIE, A.D., JONES, M.L. and ATTWOOD, T.K. (1996) DbBrowser: integrated access to databases worldwide. *TiBS*, **21**(5), 191.

Bioinformatics resource centres
EMMERT, D.B., STOEHR, P.J., STOESSER, G. and CAMERON, G.N (1994) The European Bioinformatics Institute (EBI). *Nucleic Acids Research*, **22**(17), 3445–3449.

2.13 Web addresses

WWW	http://www.w3.org/
CERN	http://www.cern.ch/
Lynx	http://lynx.browser.org
ACS	http://www.ukans.edu/~acs/
Mosaic	http://www.ncsa.uiuc.edu/SDG/Software/Mosaic/NCSA MosaicHome.html
NCSA	http://www.ncsa.uiuc.edu/ncsa.html
Netscape	http://www.netscape.com/
EMBnet	http://www.embnet.org/
Hinxton Hall	http://www.ebi.ac.uk/hinxton/hinxton.html
Sanger Centre	http://www.sanger.ac.uk/Info/
HGMP-RC	http://www.hgmp.mrc.ac.uk/HGMP.html
EBI	http://www.ebi.ac.uk/ebi_home.html
EMBL	http://www.embl-heidelberg.de/
ExPASy	http://expasy.hcuge.ch/
MIPS	http://www.mips.biochem.mpg.de/
UCL	http://www.biochem.ucl.ac.uk/bsm/dbbrowser/
SRS	http://srs.ebi.ac.uk/
NCBI	http://www.ncbi.nlm.nih.gov/
Entrez	http://www.ncbi.nlm.nih.gov/Entrez/
Virtual Tourist	http://www.vtourist.com/webmap/
ANGIS	http://www.angis.su.oz.au/
City.Net	http://www.city.net/

CHAPTER THREE

Protein information resources

3.1 Introduction

The aim of this chapter is to provide an introduction to a range of biological databases, highlighting the distinction between different data types and indicating where some of the most important resources are maintained. The chapter discusses primary sequence databases, the reasons for the development of composite sequence databases, and the emergence of a variety of secondary and tertiary pattern databases. Two structure classification resources are also briefly mentioned. Reminders of important definitions, and of the single- and three-letter amino acid codes, are given. Web addresses are provided in a separate table.

3.2 Biological databases

In Chapter 1, we saw that current efforts to sequence the entire genomes of a variety of organisms have given rise to a protein sequence/structure deficit, essentially because it is much easier rapidly to produce vast amounts of sequence information than it is to determine protein 3D structures in atomic detail. If we are to derive the maximum benefit from the deluge of sequence information, we must deal with it in a concerted way: this means establishing, maintaining and disseminating databases; providing easy-to-use software to access the information they contain; and designing state-of-the art analysis tools to visualise and interpret the structural and functional clues latent in the data.

The first step, then, in analysing sequence information is to assemble it into central, shareable resources, i.e. databases. Databases are effectively electronic filing cabinets, a convenient and efficient method of storing vast amounts of information. There are many different database types, depending both on the nature of the information being stored (e.g., sequences or structures, 2D gel or 3D structure images, and so on) and on the manner of data storage (e.g., whether in **flat-files**, tables in a **relational database**, or objects in an **object-oriented database**). Here we are concerned only with the different types of biological data, rather than on particular storage or management mechanisms.

In the context of protein sequence analysis, we will encounter primary, composite and secondary databases. Such resources store different levels of information in totally different formats. In the past, this has led to a variety of communication problems, but emerging computer technologies are beginning to provide solutions, allowing seamless, transparent access to disparate, distributed data structures over the Internet.

Primary and secondary databases are used to address different aspects of sequence analysis, because they store different levels of protein sequence information (see Box 3.1). It is therefore essential to know when and how to use them to build the most effective sequence analysis strategies. Some of the most important resources (from both current and historical viewpoints) are outlined in the following sections.

BOX 3.1: LEVELS OF PROTEIN SEQUENCE AND STRUCTURAL ORGANISATION

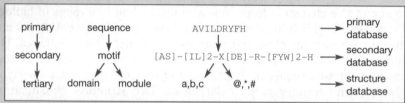

The primary structure of a protein is its amino acid sequence; these are stored in primary databases as linear alphabets that denote the constituent residues (see Box 3.2). The secondary structure of a protein corresponds to regions of local regularity (e.g., α-helices and β-strands), which, in sequence alignments, are often apparent as well-conserved motifs; these are stored in secondary databases as patterns (e.g., regular expressions, fingerprints, blocks, profiles, etc.). The tertiary structure of a protein arises from the packing of its secondary structure elements, which may form discrete domains within a fold (a,b,c), or may give rise to autonomous folding units or modules (@,*,#); complete folds, domains and modules are stored in structure databases as sets of atomic co-ordinates.

3.3 Primary sequence databases

In the early 1980s, sequence information started to become more abundant in the scientific literature. Realising this, several laboratories saw that there might be advantages to harvesting and storing these sequences in central repositories. Thus, several **primary database** projects began to evolve in different parts of the world. Table 3.1 lists some of the most important nucleic

Table 3.1 Primary nucleic acid and protein sequence databases.

Nucleic acid	Protein
EMBL	PIR
GenBank	MIPS
DDBJ	SWISS-PROT
	TrEMBL
	NRL-3D

acid and protein sequence databases that arose from such initiatives. The databanks are described briefly below.

3.3.1 Nucleic acid sequence databases

As shown in Table 3.1, the principal DNA sequence databases are GenBank (USA), EMBL (Europe) and DDBJ (Japan), which exchange data on a daily basis to ensure comprehensive coverage at each of the sites. Further details of these resources, and the structure of their entries (taking specific examples from GenBank), are discussed in Chapter 4. In the remainder of this chapter, we will be concerned with the different varieties of protein sequence and pattern database.

3.3.2 Protein sequence databases

PIR

The Protein Sequence Database was developed at the National Biomedical Research Foundation (NBRF) in the early 1960s by Margaret Dayhoff as a collection of sequences for investigating evolutionary relationships among proteins. Since 1988, the Protein Sequence Database has been maintained collaboratively by PIR-International (Barker *et al.*, 1998), an association of macromolecular sequence data collection centres: the consortium includes the Protein Information Resource (PIR) at the NBRF, the International Protein Information Database of Japan (JIPID), and the Martinsried Institute for Protein Sequences (MIPS).

In its current form, the database is split into four distinct sections, designated PIR1–PIR4, which differ in terms of the quality of data and level of annotation provided: PIR1 contains fully classified and annotated entries; PIR2 includes preliminary entries, which have not been thoroughly reviewed and may contain redundancy; PIR3 contains unverified entries, which have not been reviewed; and PIR4 entries fall into one of four categories: (i) **conceptual translations** of artefactual sequences; (ii) conceptual translations of sequences that are not transcribed or translated; (iii) protein sequences or conceptual translations that are extensively genetically engineered; or (iv) sequences that are not genetically encoded and not produced on ribosomes. Programs are provided for data retrieval and sequence searching via the NBRF-PIR database Web page.

MIPS

The Martinsried Institute for Protein Sequences collects and processes sequence data for the tripartite PIR-International Protein Sequence Database project (Mewes *et al.*, 1998). The database is distributed with PATCHX, a supplement of unverified protein sequences from external sources. Access to the database is provided through its Web server: results of FastA similarity searches of all proteins within PIR-International and PATCHX are stored in a dynamically maintained database, allowing instant access to FastA results.

SWISS-PROT

SWISS-PROT is a protein sequence database which, from its inception in 1986, was produced collaboratively by the Department of Medical Biochemistry at the University of Geneva and the EMBL; after 1994, the collaboration moved to EMBL's UK outstation, the EBI (Bairoch and Apweiler, 1998). In April 1998, further change saw a move to the Swiss Institute of Bioinformatics (SIB); hence the database is now maintained collaboratively by SIB and EBI/EMBL. The database endeavours to provide high-level annotations, including descriptions of the function of the protein, and of the structure of its domains, its post-translational modifications, variants, and so on. SWISS-PROT aims to be minimally redundant, and is interlinked to many other resources. In 1996, a computer-annotated supplement to SWISS-PROT was created, termed TrEMBL, which is described in more detail below. First, however, we will take a close look at the structure of SWISS-PROT entries.

The structure of SWISS-PROT entries

The structure of the database, and the quality of its annotations, sets SWISS-PROT apart from other protein sequence resources and has made it the database of choice for most research purposes. By mid-1998, the database contained ~70 000 entries from more than 5000 different species, the bulk of these coming from just a small number of model organisms (e.g., *Homo sapiens, Saccharomyces cerevisiae, Escherichia coli, Mus musculus* and *Rattus norvegicus*).

An example entry is shown in Figure 3.1. Each line is flagged with a two-letter code, which helps to present the information in a structured way. Entries begin with an identification (ID) line and end with a // terminator. Here, the ID line informs us that the entry name is OPSD_SHEEP, a protein with 348 amino acids. ID codes in SWISS-PROT have been designed to be informative and people-friendly; they take the form PROTEIN_ SOURCE, where the PROTEIN part of the code is an acronym that denotes the type of protein, and SOURCE indicates the organism name. The protein in this example is clearly derived from sheep and, with the eye of experience, we can deduce that it is a rhodopsin.

Unfortunately, ID codes can sometimes change, so an additional identifier, an **accession number**, is also provided, which ought to remain static between database releases. The accession number is provided on the AC

```
ID   OPSD_SHEEP     STANDARD;      PRT;   348 AA.
AC   P02700;
DT   21-JUL-1986 (REL. 01, CREATED)
DT   01-FEB-1991 (REL. 17, LAST SEQUENCE UPDATE)
DT   01-NOV-1997 (REL. 35, LAST ANNOTATION UPDATE)
DE   RHODOPSIN.
GN   RHO.
OS   OVIS ARIES (SHEEP).
OC   EUKARYOTA; METAZOA; CHORDATA; VERTEBRATA; TETRAPODA; MAMMALIA;
OC   EUTHERIA; ARTIODACTYLA.
RN   [1]
RP   SEQUENCE.
RA   PAPPIN D.J.C., ELIPOULOS E., BRETT M., FINDLAY J.B.C.;
RL   INT. J. BIOL. MACROMOL. 6:73-76(1984).
.
.
.
RN   [4]
RP   RETINAL BINDING SITE.
RX   MEDLINE; 84178280. [NCBI, Geneva]
RA   PAPPIN D.J.C., FINDLAY J.B.C.;
RL   BIOCHEM. J. 217:605-613(1984).
CC   -!- FUNCTION: VISUAL PIGMENTS ARE THE LIGHT-ABSORBING MOLECULES THAT
CC       MEDIATE VISION. THEY CONSIST OF AN APOPROTEIN, OPSIN, COVALENTLY
CC       LINKED TO CIS-RETINAL.
CC   -!- THIS RHODOPSIN HAS AN ABSORPTION MAXIMA AT 495 NM.
CC   -!- PTM: SOME OR ALL OF THE CARBOXYL-TERMINAL SER OR THR RESIDUES MAY
CC       BE PHOSPHORYLATED.
CC   -!- TISSUE SPECIFICITY: ROD SHAPED PHOTORECEPTOR CELLS WHICH MEDIATES
CC       VISION IN DIM LIGHT.
CC   -!- SUBCELLULAR LOCATION: INTEGRAL MEMBRANE PROTEIN.
CC   -!- SIMILARITY: BELONGS TO FAMILY 1 OF G-PROTEIN COUPLED RECEPTORS.
CC       BELONGS TO THE OPSIN SUBFAMILY.
DR   PIR; A30407; OOSH.
DR   GCRDB; GCR_0194; -.
DR   PROSITE; PS00237; G_PROTEIN_RECEPTOR.
DR   PROSITE; PS00238; OPSIN.
DR   PRODOM [Domain structure / List of seq. sharing at least 1 domain]
DR   SWISS-2DPAGE; GET REGION ON 2D PAGE.
DR   GPCRDB-Snakes; P02700.
KW   PHOTORECEPTOR; RETINAL PROTEIN; TRANSMEMBRANE; GLYCOPROTEIN; VISION;
KW   PHOSPHORYLATION; LIPOPROTEIN; G-PROTEIN COUPLED RECEPTOR.
FT   DOMAIN         1     36     EXTRACELLULAR.
FT   TRANSMEM      37     61     1 (POTENTIAL).
FT   DOMAIN        62     73     CYTOPLASMIC.
FT   TRANSMEM      74     98     2 (POTENTIAL).
FT   DOMAIN        99    113     EXTRACELLULAR.
.
.
.
FT   TRANSMEM     285    309     7 (POTENTIAL).
FT   DOMAIN       310    348     CYTOPLASMIC.
FT   CARBOHYD       2      2     BY SIMILARITY.
FT   CARBOHYD      15     15     BY SIMILARITY.
FT   BINDING      296    296     RETINAL CHROMOPHORE.
FT   LIPID        322    322     PALMITATE (BY SIMILARITY).
FT   LIPID        323    323     PALMITATE (BY SIMILARITY).
FT   DISULFID     110    187     BY SIMILARITY.
FT   MOD_RES      343    343     PHOSPHORYLATION (BY RK) (BY SIMILARITY).
SQ   SEQUENCE   348 AA;  38891 MW;  A3B1F1A0 CRC32;
     MNGTEGPNFY VPFSNKTGVV RSPFEAPQYY LAEPWQFSML AAYMFLLIVL GFPINFLTLY
     VTVQHKKLRT PLNYILLNLA VADLFMVGG FTTTLYTSLH GYFVFGPTGC NLEGFFATLG
     GEIALWSLVV LAIERYVVVC KPMSNFRFGE NHAIMGVAFT WVMALACAAP PLVGWSRYIP
     QGMQCSCGAL YFTLKPEINN ESFVIYMFVV HFSIPLIVIF FCYGQLVFTV KEAAAQQQES
     ATTQKAEKEV TRMVIIMVIA FLICWLPYAG VAFYIFTHQG SDFGPIFMTI PAFFAKSSSV
     YNPVIYIMMN KQFRNCMLTT LCCGKNPLGD DEASTTVSKT ETSQVAPA
//
```

Figure 3.1 Example entry from SWISS-PROT (dotted lines denote points at which, for convenience, material has been excised).

line, here P02700, which, although relatively uninformative to the human user, is nevertheless computer readable. If several numbers appear on the same AC line, the first, or primary, accession number is the most current.

Next, the DT lines provide information about the date of entry of the sequence to the database, and details of when it was last modified. The description (DE) line, or lines, then informs us of the name, or names, by which the protein is known – here simply rhodopsin. The following lines give the gene name (GN), the organism species (OS) and organism classification (OC) within the biological kingdoms.

The next section of the database provides a list of supporting references; these can be from the literature, unpublished information submitted directly from sequencing projects, data from structural or mutagenesis studies, and so on. The database is thus an important repository of information that is difficult, or impossible, to find elsewhere.

Following the references are found comment (CC) lines. These are divided into themes, which tell us about the FUNCTION of the protein, its post-translational modifications (PTM), its TISSUE SPECIFICITY, SUB-CELLULAR LOCATION, and so on. Where such information is available, the CC lines also indicate any known SIMILARITY or affiliation to particular protein families. In this example, we learn that rhodopsin is an integral membrane 'visual' protein found in rod cells; it belongs to the opsin family and to the type 1 G-protein-coupled receptor (GPCR) superfamily.

Database cross-reference (DR) lines follow the comment field. These provide links to other biomolecular databases, including primary sources, secondary databases, specialist databases, etc. For ovine rhodopsin, we find links to the primary PIR source, to the GPCR specialist database, to the PROSITE secondary database and to the ProDom domain database.

Directly after the DR lines is found a list of relevant keywords (KW), and then a number of FT lines, which form what is known as a Feature Table. The Feature Table highlights regions of interest in the sequence, including local secondary structure (such as transmembrane domains, as seen in the figure), ligand binding sites, post-translational modifications, and so on. Each line includes a key (e.g., TRANSMEM), the location in the sequence of the feature (e.g., 37–61), and a comment, which might, for example, indicate the level of confidence of a particular annotation (e.g., POTENTIAL). For our rhodopsin example, the transmembrane domain assignments result from the application of prediction software, and, therefore, in the absence of supporting experimental 3D structural data, can only be flagged as potential.

The final section of the database entry includes the sequence itself, on the SQ lines. For efficiency of storage, the single-letter amino acid code is used (see Box 3.2), each line containing 60 residues. Sequence data in SWISS-PROT correspond to the precursor form of the protein, before post-translational processing, hence information concerning the size or molecular weight will not necessarily correspond to values for the mature protein. The extent of mature proteins or peptides may be deduced by reference to the Feature Table, which will indicate the region of a sequence that

G	Glycine	Gly		P	Proline	Pro
A	Alanine	Ala		V	Valine	Val
L	Leucine	Leu		I	Isoleucine	Ile
M	Methionine	Met		C	Cysteine	Cys
F	Phenylalanine	Phe		Y	Tyrosine	Tyr
W	Tryptophan	Trp		H	Histidine	His
K	Lysine	Lys		R	Arginine	Arg
S	Serine	Ser		T	Threonine	Thr
N	Asparagine	Asn		Q	Glutamine	Gln
D	Aspartic acid	Asp		E	Glutamic acid	Glu
B	Asparagine/Aspartate	Asx		Z	Glutamine/Glutamate	Glx

The structures and properties of the amino acids are outlined in Box 3.3.

corresponds to the signal (SIGNAL), transit (TRANSIT) or pro-peptide (PROPEP) respectively. The keys CHAIN and PEPTIDE are used to denote the location of the mature form.

The structure of SWISS-PROT makes computational access to the different information fields both straightforward and efficient – for example, query software need not search the full flat-file, but can be directed to those lines that are specific to the nature of the query. For this reason, coupled with the quality of its biological annotations, SWISS-PROT has become probably the most widely used protein sequence database in the world.

TrEMBL

TrEMBL (Translated EMBL) was created in 1996 as a computer-annotated supplement to SWISS-PROT (Bairoch and Apweiler, 1998). The database benefits from the SWISS-PROT format, and contains translations of all **coding sequences** (**CDS**) in EMBL. TrEMBL has two main sections, designated SP-TrEMBL and REM-TrEMBL: SP-TrEMBL (SWISS-PROT TrEMBL) contains entries that will eventually be incorporated into SWISS-PROT, but that have not yet been manually annotated; REM-TrEMBL contains sequences that are not destined to be included in SWISS-PROT – these include immunoglobulins and T-cell receptors, fragments of fewer than eight amino acids, synthetic sequences, patented sequences, and codon translations that do not encode real proteins. TrEMBL was designed to address the need for a well-structured SWISS-PROT-like resource that would allow very rapid access to sequence data from the genome projects, without having to compromise the quality of SWISS-PROT itself by incorporating sequences with insufficient analysis and annotation.

NRL-3D

The NRL-3D database is produced by PIR from sequences extracted from the Brookhaven Protein Databank (PDB) (Namboodiri *et al.*, 1990). The

BOX 3.3: THE STRUCTURES AND PROPERTIES OF THE AMINO ACIDS

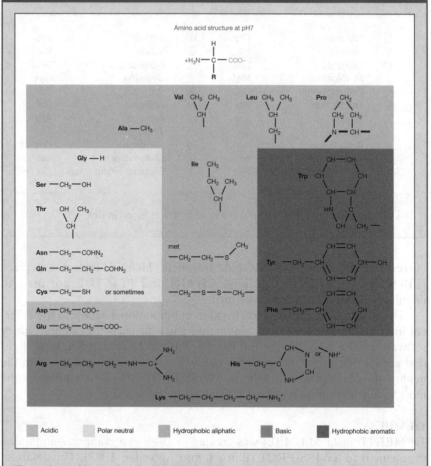

The amino acids can be divided into broad groups based on their physicochemical properties. At a superficial level, they may be classed as polar charged (acidic, basic), polar uncharged and hydrophobic (aromatic, aliphatic). But, in fact, their properties are far more complex and there is considerable overlap between them (these relationships are illustrated in Figure 8.2).

titles and biological sources of the entries conform to the nomenclature standards used in the PIR. Bibliographic references and MEDLINE cross-references are included, together with secondary structure, active site, binding site and modified site annotations, and details of experimental method, resolution, R-factor, etc. Keywords are also provided.

NRL-3D is a valuable resource, as it makes the sequence information in the PDB available both for keyword interrogation and for similarity searches. The database may be searched using the ATLAS retrieval system, a multi-database information retrieval program specifically designed to access macromolecular sequence databases.

The proliferation of primary sequence databases gives rise to a number of questions: do they all have the same format? Which is the most accurate? Which is the most up-to-date? Which is the most comprehensive? Given the choice, which should we use?

Of the protein sequence databases, NRL-3D is the least comprehensive because it reflects only the contents of the PDB, yet it has the advantage of relating directly to structural information; PIR(1–4) is the most comprehensive resource, but the quality of its annotations is still relatively poor, even in PIR1; SWISS-PROT, on the other hand, is a highly structured database (as detailed in Figure 3.1) that provides excellent annotations, but its sequence coverage is poor compared to PIR. Choosing the right database to search can seem an impossible choice; so is it, perhaps, better to search them all?

3.4 Composite protein sequence databases

One solution to the problem of proliferating primary databases is to compile a composite, i.e. a database that amalgamates a variety of different primary sources. **Composite databases** render sequence searching much more efficient, because they obviate the need to interrogate multiple resources. The interrogation process is streamlined still further if the composite has been designed to be non-redundant, as this means that the same sequence need not be searched more than once.

Different strategies can be used to create composite resources. The final product depends on the chosen data sources and the criteria used to merge them; for example, a composite resource will be non-identical if it eliminates only identical sequence copies during the amalgamation process; but if both identical and highly similar sequences are ejected (e.g., those entries that differ by only one residue, such as a leading methionine residue), then the resulting database will be more truly non-redundant.

The choice of different sources and the application of different redundancy criteria have led to the emergence of different composites (see Table 3.2), each of which has its own particular format. The main composite databases are outlined below.

3.4.1 NRDB

NRDB (Non-Redundant DataBase) is built locally at the NCBI. The database is a composite of GenPept (derived from automatic GenBank CDS translations), PDB sequences, SWISS-PROT, SPupdate (the weekly updates of SWISS-PROT), PIR and GenPeptupdate (the daily updates of GenPept). The database is thus comprehensive and contains up-to-date information. However, strictly speaking, it is not non-redundant, but non-identical, i.e., only identical sequence copies are removed from the resource. This rather simplistic approach leads to a number of problems: multiple copies of the same protein are retained in the database as a result of polymorphisms and/or minor

Table 3.2 Some of the available composite protein sequence databases, with details of their primary data sources.

NRDB	OWL	MIPSX	SP+TrEMBL
PDB	SWISS-PROT	PIR1–4	SWISS-PROT
SWISS-PROT	PIR	MIPSOwn	TrEMBL
PIR	GenBank	MIPSTrn	
GenPept	NRL-3D	MIPSH	
SWISS-PROTupdate		PIRMOD	
GenPeptupdate		NRL-3D	
		SWISS-PROT	
		EMTrans	
		GBTrans	
		Kabat	
		PseqIP	

sequencing errors; incorrect sequences that have been amended in SWISS-PROT are reintroduced when retranslated from the DNA; and numerous sequences are incorporated as full entries of existing fragments. As a result, the contents of NRDB are both error-prone and, in spite of its name, redundant. NRDB is the default database of the NCBI BLAST service.

3.4.2 OWL

OWL is a non-redundant protein sequence database built at the University of Leeds in collaboration with the Daresbury Laboratory in Warrington (Bleasby *et al.*, 1994). The database is a composite of four major primary sources: SWISS-PROT, PIR1–4, GenBank (CDS translations) and NRL-3D. The sources are assigned a priority with regard to their level of annotation and sequence validation; SWISS-PROT has the highest priority, so all others are compared against it during the amalgamation procedure. This process eliminates both identical copies of sequences and those containing single amino acid differences, leading to a compact (and efficient) resource for sequence comparisons. Nevertheless, the database suffers from many of the same problems as NRDB, which means that some sequencing errors and retranslations of incorrect sequences in GenBank are retained; and since OWL is only released on a 6–8 weekly basis, it suffers the further drawback of not being up-to-date. BLAST services for OWL are available from the UK EMBnet National Node, SEQNET, and from the UCL Specialist Node.

3.4.3 MIPSX

MIPSX is a merged database produced at the Max-Planck Institut in Martinsried (Mewes *et al.*, 1998). The database contains information from the following resources: PIR1–4; MIPS preliminary entries, MIPSOwn; MIPS/PIR preliminary entries, PIRMOD; MIPS preliminary translations, an automatic translation of EMBL; GBTrans, translated GenBank entries; Kabat; and PSeqIP. The sources are assigned a priority as denoted by their order in Table 3.2, and sequences that are identical either within or between them are removed, leaving only unique copies. In addition, all subsequences (i.e., sequences completely contained within others) are removed.

3.4.4 SWISS-PROT+TrEMBL

At the EBI, the combination of SWISS-PROT and TrEMBL provides a resource that is both comprehensive and 'minimally' redundant (Bairoch and Apweiler, 1998). This database has the advantage of containing fewer errors than do those mentioned above, yet it is still not truly non-redundant (in mid-1997, it was estimated that around 30% of the combined total of SWISS-PROT and TrEMBL was non-unique). To reduce error rates and redundancy levels further will require increasing levels of human intervention and/or the future development of expert database management systems. SWISS-PROT and TrEMBL can be searched by means of the SRS sequence retrieval system on the EBI Web server.

3.4.5 Another *embarras du choix*

Just as the proliferation of primary databases has led to difficulties in choosing the 'best' database for sequence analysis, the same difficult choices arise with the creation of several composites: which contains the highest quality data? Which is the most comprehensive? Which is the most up-to-date? Which should we use?

Ultimately, the choice of database depends on what factors are considered most important for the job in hand (and possibly on whose Web server responds most quickly!). Although not up-to-date, OWL has the advantage of being fully indexed. This means that the database has been designed for use with a query language, which allows its contents to be rapidly interrogated and manipulated in a variety of different ways. By contrast, NRDB is not available for complex querying, but is useful for up-to-date sequence searches because it contains daily updates of GenPept and weekly updates of SWISS-PROT.

Today, while Web access is still relatively easy, if sometimes a little slow, it may be better to search a number of composites, just to be quite sure that nothing obvious has been missed. Of course, this rather flies in the face of the rationale for developing composite resources, but until the truly error-free, non-redundant, comprehensive sequence database exists, this may be the only practical solution. Alternatively, it is possible to create an in-house custom composite database using the *nr* software from NCBI.

3.5 Secondary databases

In addition to the numerous primary and composite resources, there are many secondary (or pattern) databases, so-called because they contain the fruits of analyses of the sequences in the primary sources. Because there are several different primary databases, and a variety of ways of analysing protein sequences, the information housed in each of the secondary resources is different – and their formats reflect these disparities. Designing software tools that can search the different types of data, interpret the range of outputs, and assess the biological significance of the results is not a trivial task.

Although this appears to present the usual confusing picture, where nothing is consistent and there are no standards, SWISS-PROT has emerged as the most popular primary source, and many **secondary databases** now use it as their basis. Some of the main secondary resources are listed in Table 3.3.

Table 3.3 Some of the major secondary 'pattern' databases: in each case, the primary source is noted, together with the type of pattern stored. PRINTS is currently the only secondary resource to be derived from a composite.

Secondary database	Primary source	Stored information
PROSITE	SWISS-PROT	Regular expressions (patterns)
Profiles	SWISS-PROT	Weighted matrices (profiles)
PRINTS	OWL*	Aligned motifs (fingerprints)
Pfam	SWISS-PROT	Hidden Markov Models (HMMs)
BLOCKS	PROSITE/PRINTS	Aligned motifs (blocks)
IDENTIFY	BLOCKS/PRINTS	Fuzzy regular expressions (patterns)

*SWISS-PROT is OWL's highest priority source.

BOX 3.4: MOTIFS

When building a multiple alignment, as more distantly related sequences are included, insertions are often required to bring equivalent parts of adjacent sequences into the correct register. As a result of this gap insertion process, islands of conservation emerge from a backdrop of mutational change. These conserved regions (typically around 10–20 amino acids in length) tend to correspond to the core structural or functional elements of the protein. Their conserved nature allows them to be used to diagnose family membership through the application of a range of sequence analysis techniques (see Box 3.5). They are most commonly termed motifs, but are also referred to by a range of other names, such as blocks, segments or features.

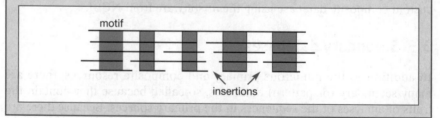

3.5.1 Why create secondary databases?

It is clear from Table 3.3 that the type of information stored in each of the secondary databases is different. Yet these resources have arisen from a common principle: namely, that homologous sequences may be gathered together in **multiple alignments**, within which are conserved regions that show little or no variation between the constituent sequences. These conserved regions, or **motifs**, usually reflect some vital biological role (i.e., are somehow crucial to the structure or function of the protein) – see Box 3.4.

Motifs have been exploited in different ways to build diagnostic patterns for particular protein families, as illustrated in Figure 3.2 (see also Box 3.5). The idea is that an unknown query sequence may be searched against a library of such patterns to determine whether or not it contains any of the

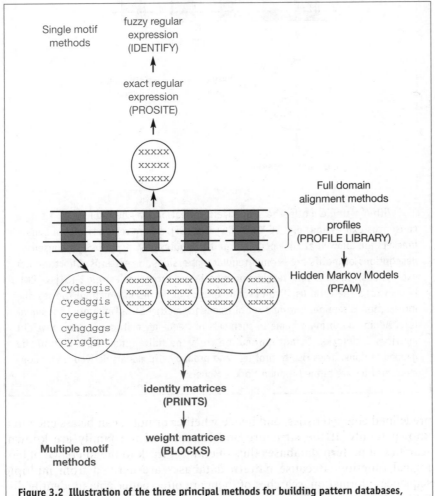

Figure 3.2 Illustration of the three principal methods for building pattern databases, i.e., using single motifs, multiple motifs and full domain alignments.

BOX 3.5: TERMS USED IN SEQUENCE ANALYSIS METHODS

At the heart of sequence analysis methods is the multiple sequence alignment. Application of these methods involves the derivation of some kind of representation, or abstraction, of conserved features of the alignment, which may be diagnostic of structure or function. Various terms are used to describe the different types of data representation, as shown in the figure.

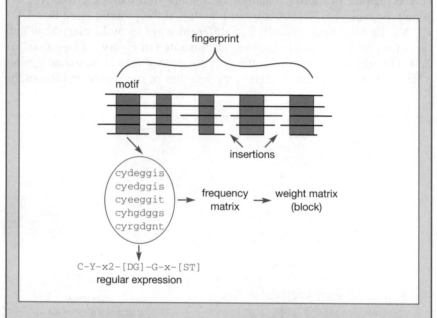

Within a single motif, the sequence information may be reduced to a single consensus expression: e.g., C-Y-x2-[DG]-G-x-[ST], where residues within square brackets are allowed at that position, and x denotes any residue. Such consensus descriptions of motifs are termed regular expressions, or patterns. Short regular expressions, three or four residues in length, are termed rules: e.g., [ST]-x-[RK]. By contrast, the term used to describe groups of motifs in which all the residue information is retained within a set of frequency matrices is a fingerprint, or signature. Adding a scoring scheme to such sets of frequency matrices results in weight matrices, or blocks. Using information from the full alignment, including the gapped regions, gives rise to profiles; and probabilistic models derived from alignment profiles are termed Hidden Markov Models.

predefined characteristics, and hence whether or not it can be assigned to a known family. If the structure and function of the family are known, searches of **pattern databases** thus offer a fast track to the inference of biological function. Because pattern databases are derived from multiple sequence information, searches of them are often better able to identify distant relationships than are corresponding searches of the primary databases. However, none of the pattern databases is yet complete; they should therefore only be used to augment primary database searches, rather than to replace them.

Some of the major secondary databases are outlined in the following pages. Details of the analysis methods that underlie their development and use are given in Chapter 8.

3.5.2 PROSITE

The first secondary database to have been developed was PROSITE, which is now maintained collaboratively at the Swiss Institute of Bioinformatics (Bairoch *et al.*, 1997). The rationale behind its development was that protein families could be simply and effectively characterised by the single most conserved motif observable in a multiple alignment of known homologues, such motifs usually encoding key biological functions (e.g., enzyme active sites, ligand or metal binding sites, etc.). Searching such a database should, in principle, help to determine to which family of proteins a new sequence might belong, or which domain(s) or functional site(s) it might contain.

Within PROSITE, motifs are encoded as **regular expressions**, often simply referred to as **patterns**. The process used to derive patterns involves the construction of a multiple alignment and manual inspection to identify conserved regions. Sequence information within individual motifs is reduced to single consensus expressions, and the resulting seed patterns are used to search SWISS-PROT. Results are checked manually to determine how well the patterns have performed: ideally, there should be only correct matches (so-called **true-positives**), and no incorrect matches (**false-positives**) – see Box 3.6. Patterns whose diagnostic performance is compromised by matching too many false-positives are fine-tuned, and SWISS-PROT is re-scanned. This process of adjustment is repeated until an optimal pattern is created.

Sometimes, a complete protein family cannot be characterised effectively by a single motif. In these cases, additional patterns are designed to encode other well-conserved parts of the alignment; the iterative fine-tuning process is then repeated until a set of patterns is achieved that is capable of capturing all, or most, of the characterised family from the given version of SWISS-PROT without matching too many, or any, false-positives.

The structure of PROSITE entries
Entries are deposited in PROSITE in two distinct files. The first of these houses the pattern and lists all matches in the parent version of SWISS-PROT; as shown in Figure 3.3, the data are structured in a manner reminiscent of SWISS-PROT entries, where each field relates to a specific type of information. The second is a documentation file, which provides details of the characterised family and, where known, a description of the biological role of the chosen motif(s) and a supporting bibliography; as shown in Figure 3.4, this is a free-format text file.

The structure of the data file is easy to understand. Like SWISS-PROT, each entry contains both an identifier (ID), which is usually some sort of acronym for the family, and an accession number (AC), which

BOX 3.6: DETERMINING SIGNIFICANCE OF DATABASE MATCHES

One of the aims of sequence analysis is to design computational methods (e.g., using discriminatory patterns for particular protein families) that help to assign functional and/or structural information to uncharacterised sequences; this is achieved by means of primary database searches, the goal of which is to identify relationships with already known sequences. Within a database, the challenge is to establish which sequences are related (these are true-positive), and which are unrelated (true-negative). In any database search, at a given scoring threshold, it is likely that several unrelated sequences will match the search pattern erroneously (so-called false-positives), and several correct matches will fail completely to be diagnosed – these are the false-negatives.

In deriving discriminatory patterns for protein families, one of the central problems is to be able to improve diagnostic performance, to capture all (or the majority of) true-positive family members, to include no (or few) false-positives, and hence minimise or preclude false-negatives. This essentially requires the small true-positive curve to be pulled apart from the larger true-negative curve depicted in the diagram, such that the overlap between them is as small as possible, or preferably eliminated – this is important because, for sequences falling within the overlapping area, it can be difficult or impossible to determine which matches are really correct (statistical approaches are commonly used to assign confidence levels to the veracity of matches in this area, but it must be remembered that mathematical significance does not equate with biological proof). Different analytical methods have been designed in an attempt to separate such curves, to improve the resolution or diagnostic performance of database search techniques in general, and of family-specific discriminators in particular.

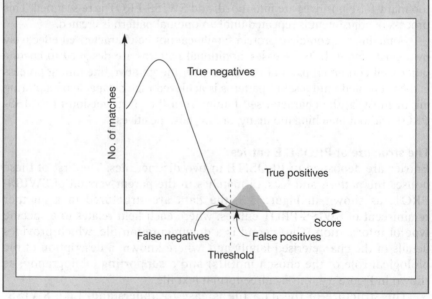

```
ID   OPSIN; PATTERN.
AC   PS00238;
DT   APR-1990 (CREATED); NOV-1997 (DATA UPDATE); NOV-1997 (INFO UPDATE).
DE   Visual pigments (opsins) retinal binding site.
PA   [LIVMW]-[PGC]-x(3)-[SAC]-K-[STALIM]-[GSACNV]-[STACP]-x(2)-[DENF]-[AP]-
PA   x(2)-[IY].
NR   /RELEASE=32,49340;
NR   /TOTAL=53(53); /POSITIVE=53(53); /UNKNOWN=0(0); /FALSE_POS=0(0);
NR   /FALSE_NEG=0; /PARTIAL=1;
CC   /TAXO-RANGE=??E??; /MAX-REPEAT=1;
CC   /SITE=5,retinal;
DR   P06002, OPS1_DROME, T; P28678, OPS1_DROPS, T; P22269, OPS1_CALVI, T;
DR   P08099, OPS2_DROME, T; P28679, OPS2_DROPS, T; P04950, OPS3_DROME, T;
DR   P28680, OPS3_DROPS, T; P08255, OPS4_DROME, T; P29404, OPS4_DROPS, T;
DR   P17646, OPS4_DROVI, T; P35362, OPSD_SPHSP, T; P41591, OPSD_ANOCA, T;
DR   P41590, OPSD_ASTFA, T; P02699, OPSD_BOVIN, T; P32308, OPSD_CANFA, T;
DR   P32309, OPSD_CARAU, T; P22328, OPSD_CHICK, T; P28681, OPSD_CRIGR, T;
DR   P08100, OPSD_HUMAN, T; P15409, OPSD_MOUSE, T; P35403, OPSD_POMMI, T;
DR   P02700, OPSD_SHEEP, T; P29403, OPSD_XENLA, T; P22671, OPSD_LAMJA, T;
DR   P31355, OPSD_RANPI, T; P24603, OPSD_LOLFO, T; P09241, OPSD_OCTDO, T;
DR   P35356, OPSD_PROCL, T; P31356, OPSD_TODPA, T; P35360, OPS1_LIMPO, T;
DR   P35361, OPS2_LIMPO, T; P32310, OPSB_CARAU, T; P28682, OPSB_CHICK, T;
DR   P35357, OPSB_GECGE, T; P03999, OPSB_HUMAN, T; P28684, OPSV_CHICK, T;
DR   P22330, OPSG_ASTFA, T; P22331, OPSH_ASTFA, T; P32311, OPSG_CARAU, T;
DR   P32312, OPSH_CARAU, T; P28683, OPSG_CHICK, T; P35358, OPSG_GECGE, T;
DR   P04001, OPSG_HUMAN, T; P41592, OPSR_ANOCA, T; P22332, OPSR_ASTFA, T;
DR   P32313, OPSR_CARAU, T; P22329, OPSR_CHICK, T; P04000, OPSR_HUMAN, T;
DR   P34989, OPSL_CALJA, T; P35359, OPSU_BRARE, T; P23820, REIS_TODPA, T;
DR   P47803, RGR_BOVIN , T; P47804, RGR_HUMAN , T;
DR   P17645, OPS3_DROVI, P;
DO   PDOC00211;
//
```

Figure 3.3 Example regular expression entry from PROSITE, showing the data file for the opsin pattern.

takes the form PS00000. The ID line also indicates the type of **discriminator** to expect in the file – the word PATTERN here tells us to expect a regular expression. A title, or description of the family, is contained in the DE line, and the pattern itself resides on PA lines. The following NR lines provide technical details about the derivation and **diagnostic performance** (or **diagnostic power**) of the pattern (for this reason, they are probably the most important lines to inspect when first viewing a PROSITE entry – large numbers of false-positive and **false-negative** results are indicative of poorly-performing patterns). In the example shown in Figure 3.3, we learn that the pattern was derived from release 32 of SWISS-PROT, which contained 49 340 sequences; it matched a total of 53 sequences, all of which are true-positives – in other words, this is a good pattern, with no false matches.

The comment (CC) lines provide information on the taxonomic range of the family (defined here as **eukaryotes**), the maximum number of observed repeats of the pattern (here just one), functional site annotations

```
{PDOC00211}
{PS00238; OPSIN}
{BEGIN}
*******************************************************
* Visual pigments (opsins) retinal binding site *
*******************************************************
```

Visual pigments [1,2] are the light-absorbing molecules that mediate vision. They consist of an apoprotein, opsin, covalently linked to the chromophore cis-retinal. Vision is effected through the absorption of a photon by cis-retinal which is isomerized to trans-retinal. This isomerization leads to a change of conformation of the protein. Opsins are integral membrane proteins with seven transmembrane regions that belong to family 1 of G-protein coupled receptors (see <PDOC00210>).

In vertebrates four different pigments are generally found. Rod cells, which mediate vision in dim light, contain the pigment rhodopsin. Cone cells, which function in bright light, are responsible for color vision and contain three or more color pigments (for example, in mammals: red, blue and green).

In Drosophila, the eye is composed of 800 facets or ommatidia. Each ommatidium contains eight photoreceptor cells (R1-R8): the R1 to R6 cells are outer cells, R7 and R8 inner cells. Each of the three types of cells (R1-R6, R7 and R8) expresses a specific opsin.

Proteins evolutionary related to opsins include squid retinochrome, also known as retinal photoisomerase, which converts various isomers of retinal into 11-cis retinal and mammalian retinal pigment epithelium (RPE) RGR [3], a protein that may also act in retinal isomerization.

The attachment site for retinal in the above proteins is a conserved lysine residue in the middle of the seventh transmembrane helix. The pattern we developed includes this residue.

```
-Consensus pattern:  [LIVMW]-[PGC]-x(3)-[SAC]-K-[STALIM]-[GSACNV]-
                     [STACP]-x(2)-[DENF]-[AP]-x(2)-[IY]
                     [K is the retinal binding site]
```
-Sequences known to belong to this class detected by the pattern: ALL.
-Other sequence(s) detected in SWISS-PROT: NONE.
-Last update: November 1997 / Pattern and text revised.

[1] Applebury M.L., Hargrave P.A.
     Vision Res. 26:1881-1895(1986).
[2] Fryxell K.J., Meyerowitz E.M.
     J. Mol. Evol. 33:367-378(1991).
[3] Shen D., Jiang M., Hao W., Tao L., Salazar M., Fong H.K.W.
     Biochemistry 33:13117-13125(1994).
{END}

Figure 3.4 Example regular expression entry from PROSITE, showing the documentation file for the opsin pattern shown in Figure 3.3.

(in this example, the retinal binding site is encoded in the fifth position of the pattern), and so on. Following the comments are lists of the accession numbers and SWISS-PROT identification codes of all the true matches to the pattern (denoted by T), and any 'possible' matches (denoted by P), which are often fragments. Although there are no false-positive or false-negative matches in this example, when these do occur, they are listed and are denoted by the letters F and N respectively (the number of false and missed matches is also documented in the NR lines). The final line of the file (DO) points to the associated family documentation file.

The structure of the documentation file is much simpler. Each entry is identified by its own accession number (which takes the form PDOC00000), and provides a cross-reference to the accession number and identifier of its data file. A free-format description of the family then follows, including details of the pattern and, where known, its biological relevance. The file concludes with appropriate bibliographic references, as shown in Figure 3.4. The database is accessible for keyword and sequence searching via the ExPASy Web server.

3.5.3 PRINTS

From inspection of sequence alignments, it is clear that most protein families are characterised not by one, but by several conserved motifs. It therefore makes sense to use many, or all, of these to build diagnostic signatures of family membership. This is the principle behind the development of the PRINTS **fingerprint** database, which until 1999 was maintained in the Department of Biochemistry and Molecular Biology at University College London (UCL) (Attwood *et al.*, 1998). Fingerprints inherently offer improved diagnostic reliability over single-motif methods by virtue of the mutual context provided by motif neighbours: in other words, if a query sequence fails to match all the motifs in a given fingerprint, the pattern of matches formed by the remaining motifs still allows the user to make a reasonably confident diagnosis.

Within PRINTS, motifs are encoded as ungapped, unweighted local alignments. The process used to derive fingerprints differs markedly from that used to create regular expressions. Here, sequence information in a set of seed motifs is augmented through a process of iterative (composite) database scanning. In brief, from a small initial multiple alignment, conserved motifs are identified and excised manually for database searching (PRINTS is currently derived from scans of OWL, but future releases will be built from searches of SWISS-PROT + SP-TrEMBL). Results are examined to determine which sequences have matched all the motifs within the fingerprint; if there are more matches than were in the initial alignment, the additional information from these new sequences is added to the motifs, and the database is searched again. This iterative process is repeated until no further complete fingerprint matches can be identified. The results are then annotated for inclusion in the database.

Figure 3.5 illustrates three different aspects of a PRINTS entry. At the top of the file (Figure 3.5(a)), each fingerprint is given an identifying code

(a)

```
OPSIN           OPSIN SIGNATURE
Type of fingerprint: COMPOUND with 3 elements
Links:
    PRINTS; PR00237 GPCRRHODOPSN; PR00247 GPCRCAMP; PR00248 GPCRMGR
    PRINTS; PR00249 GPCRSECRETIN; PR00250 GPCRSTE2; PR00251 BACTRLOPSIN
    PROSITE; PS00238 OPSIN; PS00237 G_PROTEIN_RECEPTOR
    BLOCKS; BL00238
    SBASE; OPSD_HUMAN
    GCRDB; GCR_0085
    Creation date 20-DEC-1993; UPDATE 2-JUL-1996

    1. APPLEBURY, M.L. and HARGRAVE, P.A.
    Molecular biology of the visual pigments.
    VISION RES. 26 (12) 1881-1895 (1986).
```

(b)

```
SUMMARY INFORMATION
  73 codes involving 3 elements
   1 codes involving 2 elements

COMPOSITE FINGERPRINT INDEX
   3|  73   73   73
   2|   0    1    1
  --+----------------
    |   1    2    3
```

(c)

```
INITIAL MOTIF SETS
OPSIN1 Length of motif = 13 Motif number = 1
Opsin motif I - 1
                     PCODE          ST    INT
YVTVQHKKLRTPL        OPSD_BOVIN     60    60
YVTVQHKKLRTPL        OPSD_HUMAN     60    60
YVTVQHKKLRTPL        OPSD_SHEEP     60    60
AATMKFKKLRHPL        OPSG_HUMAN     76    76
AATMKFKKLRHPL        OPSR_HUMAN     76    76
YIFATTKSLRTPA        OPS1_DROME     73    73
VATLRYKKLRQPL        OPSB_HUMAN     57    57
YIFGGTKSLRTPA        OPS2_DROME     80    80
WVFSAAKSLRTPS        OPS3_DROME     81    81
WIFSTSKSLRTPS        OPS4_DROME     77    77
YLFSKTKSLQTPA        OPSD_OCTDO     58    58
YLFTKTKSLQTPA        OPSD_LOLFO     57    57

OPSIN2 Length of motif = 13 Motif number = 2
Opsin motif II - 1
                     PCODE          ST    INT
GWSRYIPEGMQCS        OPSD_BOVIN    174    101
GWSRYIPEGLQCS        OPSD_HUMAN    174    101
GWSRYIPQGMQCS        OPSD_SHEEP    174    101
GWSRYWPHGLKTS        OPSG_HUMAN    190    101
GWSRYWPHGLKTS        OPSR_HUMAN    190    101
GWSRYVPEGNLTS        OPS1_DROME    187    101
GWSRFIPEGLQCS        OPSB_HUMAN    171    101
GWSAYVPEGNLTA        OPS2_DROME    194    101
TWGRFVPEGYLTS        OPS3_DROME    194    100
FWDRFVPEGYLTS        OPS4_DROME    190    100
NWGAYVPEGILTS        OPSD_OCTDO    174    103
GWGAYTLEGVLCN        OPSD_LOLFO    173    103
```

Figure 3.5 Excerpt from the PRINTS database, illustrating three different aspects of an entry, showing: (a) the general ID code and database cross-links; (b) a summary of the diagnostic performance of the fingerprint; and (c) how fingerprints are stored as ungapped aligned motifs.

(usually an acronym that attempts to describe the family), and a title that gives the family name – here, the fingerprint, or signature, for the opsins is identified by the code OPSIN. All entries also have unique accession numbers, which take the form PR00000 (not shown). An indication is then given of the number of motifs in the fingerprint (here 3). Prior to the date line, which indicates when the entry was added to the database and when it was last updated, a number of database cross-links are provided, allowing users to access additional information about the family in related biological resources. The final parts of this initial section provide bibliographic information, and a brief description of the characterised family, coupled with technical details concerning the derivation of the fingerprint (not shown). Where possible, the description includes details of the structural and/or functional relevance of the conserved motifs.

In the second section of the PRINTS entry, Figure 3.5(b), is found information relating to the diagnostic performance both of the fingerprint as a whole and of its constituent motifs. First, a summary lists how many sequences matched all the motifs and how many made partial matches (i.e., failed to match one or more motifs) – in this example, we learn that 73 sequences matched all three elements of the fingerprint, and one sequence matched only two motifs. The table that follows provides additional information in support of these results, detailing how many sequences were matched by each individual motif – here, the important information gained is that the reported partial hit failed to match motif 1.

In the final part of the entry, Figure 3.5(c), are listed the seed motifs used to generate the fingerprint, followed by the final motifs (not shown) that result from the iterative database scanning procedure. Each motif is identified by its parent ID code plus a number that indicates which component of the fingerprint it is: in this example, the three motifs in the OPSIN fingerprint are designated OPSIN1, OPSIN2 and OPSIN3 (motif 3 is not shown). After the code, the motif length is given, followed by a short description, which indicates the relevant iteration number (for the initial motifs, of course, this will always be 1). The aligned motifs themselves are then provided, together with the corresponding source database ID code of each of the constituent sequence fragments (here only sequences from SWISS-PROT were included in the initial alignment). The location in the parent sequence of each fragment is then given, together with the interval (i.e., the number of residues) between the fragment and its preceding neighbour – for the first motif, this value is the distance from the N-terminus.

An important consequence of storing the motifs in this 'raw' form is that, unlike with regular expressions or other such abstractions, no sequence information is lost. This means that a variety of different scoring methods may be superposed onto the motifs, providing different scoring potentials for, and hence different perspectives on, the same underlying data. PRINTS may therefore provide the raw material for automatically derived **tertiary databases**.

The database is accessible for keyword and sequence searching via the DbBrowser Bioinformatics Web server, which in 1999 will have relocated from UCL to the University of Manchester.

PROSITE and PRINTS are set apart from other secondary databases by virtue of the manually input family documentations, which help to place conserved sequence information in a structural or functional context. This is vital for the end user, who not only wants to discover, for example, whether a novel sequence has matched a predefined sequence motif, but, more importantly, also needs to understand its biological significance.

The following sections briefly describe some related secondary and tertiary databases that are generated using more automated procedures and, as a consequence, provide little or no family annotation. Some of these use PRINTS and PROSITE as their data sources.

3.5.4 BLOCKS

The diagnostic limitations of regular expressions led to the creation of a multiple-motif database, based on protein families contained in PROSITE, at the Fred Hutchinson Cancer Research Center (FHCRC) in Seattle; this is the BLOCKS database (Henikoff *et al.*, 1998). In this resource, the motifs, or **blocks**, are created by automatically detecting the most highly conserved regions of each protein family: this is achieved via a method based, in its initial stages, on the identification of three conserved amino acids (which need not be contiguous in sequence). The resulting blocks, which are ultimately encoded as ungapped local alignments, are calibrated against SWISS-PROT to obtain a measure of the likelihood of a chance match. Two scores are noted for each block: the first denotes the level at which 99.5% of matches are **true-negatives;** the second is the median value of the true-positive scores, which, for the purpose of comparing the diagnostic performance of individual blocks, is normalised by multiplying by 1000 and dividing by the 99.5% score. The median standardised score for known true-positive matches is termed strength.

The structure of BLOCKS entries

Figure 3.6 illustrates a typical block. The structure of the database entry is compatible with that used in PROSITE, where each block is identified by a general code (the ID line) and an accession number, which takes the form BL00000X (X is a letter that specifies which the block is within the family's set of blocks, e.g., BL00327C is the third bacterial rhodopsin block). Similarly, the ID line indicates the type of discriminator to expect in the file – here, not surprisingly, the word BLOCK tells us to expect a block. The AC line also provides an indication of the minimum and maximum distances of the block from its preceding neighbour, or from the N-terminus if it is the first in a set of blocks. A title, or description of the family, is contained in the DE line. This is followed by the BL line, which provides an indication of the diagnostic power and some physical details of the block: these include the amino acid triplet (here R-Y-A), the width of the block and the number of sequences it contains, the 99.5%-level score, and finally the strength. Strong blocks are more effective than weak blocks (strength less than 1100) at separating true-positives from true-negatives. Following this

```
ID    BACTERIAL_OPSIN_RET; BLOCK
AC    BL00327C; distance from previous block=(2,4)
DE    Bacterial rhodopsins retinal binding site proteins.
BL    RYA motif; width=29; seqs=6; 99.5%=1206; strength=1504
BAC1_HALS1 ( 144) LARYTWWLFSTICMIVVLYFLATSLRAAA 61

BAC2_HALS2 ( 143) LARYTWWLFSTIAFLFVLYYLLTSLRSAA 52

BACR_HALHA ( 145) SYRFVWWAISTAAMLYILYVLFFGFTSKA 73

BACT_NATPH ( 121) IERYALFGMGAVAFLGLVYYLVGPMTESA 100

BACH_HALSS ( 164) LLRWVWYAISCAFFVVVLYILLAEWAEDA 59
BACH_NATPH ( 174) LMRWFWYAISCACFLVVLYILLVEWAQDA 56
//
```

Figure 3.6 Example entry from BLOCKS, showing the third block used to characterise the bacterial rhodopsins.

information comes the block itself, which indicates the SWISS-PROT IDs of the constituent sequences, the start position of the fragment, the sequence fragment itself, and a score, or weight, that provides a measure of the closeness of the relationship of that sequence to others in the block (100 being the most distant). Sequence fragments that are less than 80% similar are separated by blank lines.

Because the database is derived by fully automatic methods, the blocks are not annotated, but links are made to the corresponding PROSITE family documentation file. The database is accessible for keyword and sequence searching via the Blocks Web server at the FHCRC.

Blocks-format PRINTS
In addition to the BLOCKS database, the FHCRC Web server provides a version of the PRINTS database in BLOCKS format (Henikoff *et al.*, 1998). In this resource, the scoring methods that underlie the derivation of blocks have been applied to each of the aligned motifs in PRINTS. Figure 3.7 illustrates a typical motif in BLOCKS format. The structure of the entry is identical to that used in BLOCKS, with only minor differences occurring on the AC and BL lines. On the AC line, the PRINTS accession number is given, with an appended letter to indicate which component of the fingerprint it is (in the example, PR00238A indicates that this is the first motif). On the BL line, the triplet information is replaced by the word 'adapted', indicating that the motifs have been taken from another database.

Because BLOCKS-format PRINTS is derived automatically from PRINTS, its blocks are not annotated. Nevertheless, family and motif documentation may be accessed via links to the corresponding PRINTS entry. The database is accessible for keyword and sequence searching via the Blocks Web server at the FHCRC.

A further important consequence of the direct derivation of the BLOCKS databases from PROSITE and PRINTS is that they offer no

```
ID   OPSIN; BLOCK
AC   PR00238A; distance from previous block=(31,81)
DE   OPSIN SIGNATURE
BL   adapted; width=13; seqs=73; 99.5%=858; strength=1172;
OPSD_CHICK ( 60) YVTIQHKKLRTPL 11
OPSD_RANPI ( 60) YVTIQHKKLRTPL 11
   S79840  ( 60) YVTIQHKKLRTPL 11
OPSD_BOVIN ( 60) YVTVQHKKLRTPL 11
OPSD_CANFA ( 60) YVTVQHKKLRTPL 11
OPSD_CRIGR ( 60) YVTVQHKKLRTPL 11
OPSD_HUMAN ( 60) YVTVQHKKLRTPL 11
OPSD_MOUSE ( 60) YVTVQHKKLRTPL 11
OPSD_SHEEP ( 60) YVTVQHKKLRTPL 11
.

.

OPSD_OCTDO ( 58) YLFSKTKSLQTPA 35
OPSD_LOLFO ( 57) YLFTKTKSLQTPA 38
OPSD_TODPA ( 56) YLFTKTKSLQTPA 38
  ASOPSIN  ( 50) YLFTKTKSLQTPA 38
OPSB_CHICK ( 67) FCTARFRKLRSHL 69
  HMIORH2  ( 77) YLFNKSAALRTPA 100
//
```

Figure 3.7 **Example entry from BLOCKS-format PRINTS,** showing the first block used to characterise the opsin family.

further family coverage. Nevertheless, as different methods are used to construct the blocks in each of the databases, it is worth searching both. Still more, ~50% of families encoded in PRINTS are not represented in PROSITE, so searches of both BLOCKS databases will be more comprehensive than searches of either resource alone.

3.5.5 Profiles

An alternative philosophy to the motif-based approach of protein family characterisation adopts the principle that the variable regions between conserved motifs also contain valuable sequence information. Here, the complete sequence alignment effectively becomes the discriminator. The discriminator, termed a **profile**, is weighted to indicate where insertions and deletions (**INDELs**) are allowed, what types of residues are allowed at what positions, and where the most conserved regions are. Profiles (alternatively known as **weight matrices**) provide a sensitive means of detecting distant sequence relationships, where only very few residues are well conserved – in these circumstances, regular expressions cannot provide good discrimination, and will either miss too many true-positives or catch too many false ones.

The limitations of regular expressions in identifying distant homologues led to the creation of a compendium of profiles at the Swiss Institute for Experimental Cancer Research (ISREC) in Lausanne. Each profile has separate data and family-annotation files whose formats are compatible with PROSITE data and documentation files. This allows results that have been annotated to an appropriate standard to be made available as an integral part of PROSITE (Bairoch *et al.*, 1997).

The structure of PROSITE profile entries

Figure 3.8 shows an excerpt from a profile data file. The structure of the file is based on that of PROSITE, but with obvious differences. The first change is seen on the ID line, where the word MATRIX indicates that the type of discriminator to expect is a profile. Pattern (PA) lines are replaced by matrix (MA) lines, which list the various parameter specifications used to derive and describe the profile: they include details of the alphabet used (i.e., whether nucleic acid {ACGT} or amino acid {ABCDEFGHIKLMNPQRSTVWYZ}), the length of the profile, cut-off scores (which are designed, as far as possible, to exclude random matches), and so on. The I and M fields contain position-specific profile scores for insert and match positions respectively.

Profiles that have not achieved the standard of annotation necessary for inclusion in PROSITE are nevertheless made available for searching via the ISREC Web server.

```
ID   SH3; MATRIX.
AC   PS50002;
DT   NOV-1995 (CREATED); NOV-1995 (DATA UPDATE); NOV-1995 (INFO UPDATE).
DE   Src homology 3 (SH3) domain profile.
MA   /GENERAL_SPEC: ALPHABET='ACDEFGHIKLMNPQRSTVWY';
MA   /CUT_OFF: LEVEL=0; SCORE=90; N_SCORE=7.0; MODE=1;
.
.
MA   /M: SY='V';M=0,-4,-3,-4,-1,-3,-3,5,-3,3,3,-2,-2,-2,-3,-2,0,5,-8,-4;
MA   /M: SY='L';M=-1,-6,-3,-3,-1,-3,-2,2,-3,3,2,-2,-2,-2,-3,-2,-1,2,-5,-3;
MA   /M: SY='D';M=0,-6,3,3,-6,0,1,-3,2,-5,-2,2,-1,2,1,0,0,-4,-7,-5;
MA   /M: SY='K';M=-1,-6,0,0,-2,-1,0,-3,3,-4,-1,1,-1,0,1,0,0,-3,-6,-4;
MA   /M: SY='N';M=1,-4,1,1,-5,0,0,-2,0,-3,-2,1,1,0,-1,1,1,-1,-7,-5;
MA      /I: MI=0; I=-1; MD=0; /M: SY='X'; M=0; D=-1;
MA   /M: SY='G';M=1,-5,0,0,-5,1,-2,-1,-2,-3,-2,0,0,-1,-2,0,0,-1,-8,-6;
MA   /M: SY='G';M=1,-6,3,3,-7,3,0,-4,-1,-5,-4,2,-1,1,-2,1,0,-3,-10,-6;
MA   /M: SY='W';M=-9,-12,-9,-11,1,-11,-4,-8,-5,-3,-6,-6,-8,-7,3,-4,-8,-9,26,0;
MA   /M: SY='W';M=-7,-9,-9,-9,0,-9,-4,-5,-5,-1,-4,-6,-7,-6,2,-3,-6,-6,18,-1;
MA   /M: SY='K';M=-1,-7,0,0,-3,-2,0,-2,2,-3,-1,1,-1,1,2,0,-1,-3,-5,-5;
MA      /I: MI=0; I=-2; MD=0; /M: SY='X'; M=0; D=-2;
.
.
```

Figure 3.8 Excerpt from a PROSITE profile entry, illustrating part of the profile used to characterise the SH3 domain.

3.5.6 Pfam

Just as there are different ways of using motifs to characterise protein families (e.g., depending on the scoring scheme used), so there are different methods of using full sequence alignments to build family discriminators. An alternative to the use of profiles is to encode alignments in the form of **Hidden Markov Models (HMMs)**. These are statistically based mathematical treatments, consisting of linear chains of match, delete or insert states that attempt to encode the sequence conservation within aligned families.

A collection of HMMs for a range of protein domains is provided by the Pfam database, which is maintained at the Sanger Centre (Sonnhammer *et al.*, 1998). The database is based on two distinct classes of alignment: hand-edited seed alignments, which are deemed to be accurate (these are used to produce Pfam-A); and those derived by automatic clustering of SWISS-PROT, which are less reliable (these give rise to Pfam-B).

The high-quality seed alignments are used to build HMMs, to which sequences are automatically aligned to generate final full alignments. If the initial alignments do not produce diagnostically sound HMMs, the seed is improved and the gathering process iterated until a good result is achieved. The methods that ultimately generate the best full alignment may vary for different families, so the parameters are saved in order that the result can be reproduced. The collection of seed and full alignments, coupled with minimal annotations, database and literature cross-references, and the HMMs themselves, constitute Pfam-A. All sequence domains that are not included in Pfam-A are automatically clustered and deposited in Pfam-B.

The structure of Pfam entries

Figure 3.9 shows some of the information used to describe a Pfam-A entry. The format is compatible with PROSITE, each entry being identified by both an accession (AC) number (which takes the form PF00000) and an ID code (a single keyword). DE lines provide the title, or description, of the family, and AU lines indicate the author of the entry. The methods used to create both the seed and the full automatic alignment are noted on AL and AM lines respectively. The source database suggesting that seed members belong to one family, appropriate database cross-references, and the search program and cut-off used to build the full alignment are given in the SE, DR and GA lines.

```
AC  PF00001
ID  7tm
DE  7 transmembrane receptor (rhodopsin family)
AU  Sonnhammer ELL
AL  HMM_simulated_annealing
AM  hmma -qR
SE  Prosite
DR  PROSITE; PDOC00210; [Expasy][SRS Japan|UK|USA|]
DR  PROSITE; PDOC00211; [Expasy][SRS Japan|UK|USA|]
GA  Bic_raw 23.83 hmmfs 15
DR  http://www.gcrdb.uthscsa.edu/
```

Figure 3.9 Excerpt from the 7TM Pfam entry, showing information used to build the seed and full alignments.

Although entries in Pfam-A have an annotation file available (which may contain details of the method, a description of the domain, and links to other databases), for the majority of entries, extensive family annotations are not yet in place.

Pfam is accessible for sequence searching via the Web server at the Sanger Centre on the Hinxton Genome Campus.

3.5.7 IDENTIFY

Another automatically derived tertiary resource, derived from BLOCKS and PRINTS, is IDENTIFY, which is produced in the Department of Biochemistry at Stanford University (Nevill-Manning *et al.*, 1998). The program used to generate this resource, eMOTIF, is based on the generation of consensus expressions from conserved regions of sequence alignments. However, rather than encoding the exact information observed at each position in an alignment (or motif), eMOTIF adopts a 'fuzzy' approach in which alternative residues are tolerated according to a set of prescribed groupings, as illustrated in Table 3.4. These groups correspond to various biochemical properties, such as charge and size, theoretically ensuring that the resulting motifs have sensible biochemical interpretations.

Although this technique is designed to be more flexible than exact regular expression matching, its inherent permissiveness brings with it an inevitable signal-to-noise trade-off: i.e., the resulting patterns not only have the potential to make more true-positive matches, but they will consequently also match more false-positives. However, when using the resource for sequence searching, different levels of stringency are offered from which to infer the significance of matches.

IDENTIFY and its search software, eMOTIF, are accessible for use via the protein function Web server from the Biochemistry Department at Stanford.

Table 3.4 Sets of amino acids and their properties as used in eMOTIF.

Residue property	*Residue groups*
Small	Ala, Gly
Small hydroxyl	Ser, Thr
Basic	Lys, Arg
Aromatic	Phe, Tyr, Trp
Basic	His, Lys, Arg
Small hydrophobic	Val, Leu, Ile
Medium hydrophobic	Val, Leu, Ile, Met
Acidic/amide	Asp, Glu, Asn, Gln
Small/polar	Ala, Gly, Ser, Thr, Pro

While there is some overlap between them, the contents of the PROSITE, PRINTS, profiles and Pfam databases are different. In 1998, together they encode ~1500 protein families, covering a range of globular and membrane proteins, modular polypeptides, and so on. It has been estimated that the total number of protein families might be in the range 1000 to 10 000, so there is still a long way to go before any of the secondary databases can be considered to be complete. Thus, in building a search strategy, it is good practice to include all available secondary resources, to ensure both that the analysis is as comprehensive as possible and that it takes advantage of a variety of search methods.

3.6 Composite protein pattern databases

If, today, comprehensive sequence analysis requires accessing a variety of disparate databases, gathering the range of different outputs and arriving at some sort of consensus view of the results, in the future, secondary database searching will undoubtedly become more straightforward. The curators of PROSITE, Profiles, PRINTS and Pfam are now co-operating with a view to creating a unified database of protein families. The aim is to provide a single, central family annotation resource in Geneva (based on existing documentation in PROSITE and PRINTS), each entry in which will point to different discriminators in the parent PROSITE, Profiles, PRINTS or Pfam databases. This will simplify sequence analysis for the user, who will thereby have access to a one-stop-shop for protein family diagnosis.

This effort is also supported by the curators of the BLOCKS databases, who, realising the problems associated with providing detailed family documentation, are developing a dedicated protein family Web site, termed proWeb (Henikoff *et al.*, 1996). This facility provides information about individual families through hyperlinks to existing Web resources maintained by researchers in their own fields. The curators of proWeb see its primary utility as being similar to that of written reviews, but with the advantage that it can be readily updated and can include, for example, animated material, which is beyond the scope of print technology. ProWeb will greatly facilitate the task of secondary database annotators, by providing convenient access to family information and obviating the need for annotators themselves to become 'expert' on all proteins.

3.7 Structure classification databases

A chapter concerning the repertoire of biological databases that may be used to aid sequence analysis would not be complete without some consideration of protein structure classification resources. Of course, these are currently limited to the relatively few 3D structures available from crystallographic and spectroscopic studies, but their impact will increase as more structures become available.

common evolutionary origins. The evolutionary process involves substitutions, insertions and deletions in amino acid sequences. For distantly related proteins, such changes can be extensive, yielding folds in which the numbers and orientations of secondary structures vary considerably. However, where, for example, the functions of proteins are conserved, the structural environments of critical active site residues are also conserved. In an attempt to better understand sequence/structure relationships and the underlying evolutionary processes that give rise to different fold families, a variety of structure classification schemes have been established.

The nature of the information presented by a structure classification scheme is entirely dependent on the underlying philosophy of the approach, and hence on the methods used to identify and evaluate structural similarity. Structural families derived, for example, using algorithms that search and cluster on the basis of common motifs will be different from those generated by procedures based on global structure comparison; and the results of such automatic procedures will differ again from those based on visual inspection, where software tools are used essentially to render the task of classification more manageable.

Two well-known classification schemes are outlined below.

3.7.1 SCOP

The SCOP (Structural Classification of Proteins) database maintained at the MRC Laboratory of Molecular Biology and Centre for Protein Engineering describes structural and evolutionary relationships between proteins of known structure (Murzin *et al.*, 1995). Because current automatic structure comparison tools cannot reliably identify all such relationships, SCOP has been constructed using a combination of manual inspection and automated methods. The task is complicated by the fact that protein structures show such variety, ranging from small, single domains to vast multi-domain assemblies. In some cases (e.g., some modular proteins), it may be meaningful to discuss a protein structure at the same time both at the multi-domain level and at the level of its individual domains.

SCOP classification
Proteins are classified in a hierarchical fashion to reflect their structural and evolutionary relatedness. Within the hierarchy there are many levels, but principally these describe the family, superfamily and fold. The boundaries between these levels may be subjective, but the higher levels generally reflect the clearest structural similarities.

- *Family.* Proteins are clustered into families with clear evolutionary relationships if they have sequence identities $\geqslant 30\%$. But this is not an absolute measure – in some cases (e.g., the globins), it is possible to infer common descent from similar structures and functions in the absence of significant sequence identity (some members of the globin family share only 15% identity).

- **Superfamily.** Proteins are placed in superfamilies when, in spite of low sequence identity, their structural and functional characteristics suggest a common evolutionary origin.

- **Fold.** Proteins are classed as having a common fold if they have the same major secondary structures in the same arrangement and with the same topology, whether or not they have a common evolutionary origin. In these cases, the structural similarities could have arisen as a result of physical principles that favour particular packing arrangements and fold topologies.

SCOP is accessible for keyword interrogation via the MRC Laboratory Web server.

3.7.2 CATH

The CATH (Class, Architecture, Topology, Homology) database is a hierarchical domain classification of protein structures maintained at UCL (Orengo *et al.*, 1997). The resource is largely derived using automatic methods, but manual inspection is necessary where automatic methods fail. Different categories within the classification are identified by means of both unique numbers (by analogy with the **enzyme classification** or **E.C. system** for enzymes) and descriptive names. Such a numbering scheme allows efficient computational manipulation of the data. There are five levels within the hierarchy:

- **Class** is derived from gross secondary structure content and packing. Four classes of domain are recognised: (i) mainly-α, (ii) mainly-β, (iii) α–β, which includes both alternating α/β and $\alpha+\beta$ structures, and (iv) those with low secondary structure content.

- **Architecture** describes the gross arrangement of secondary structures, ignoring their connectivities; it is currently assigned manually using simple descriptions of the secondary structure arrangements (e.g., barrel, roll, sandwich, etc.).

- **Topology** gives a description that encompasses both the overall shape and the connectivity of secondary structures. This is achieved by means of structure comparison algorithms that use empirically derived parameters to cluster the domains. Structures in which at least 60% of the larger protein matches the smaller are assigned to the same topology level.

- **Homology** groups domains that share $\geqslant 35\%$ sequence identity and are thought to share a common ancestor, i.e. are homologous. Similarities are first identified by sequence comparison and subsequently by means of a structure comparison algorithm.

- **Sequence** provides the final level within the hierarchy, whereby structures within homology groups are further clustered on the basis of sequence identity. At this level, domains have sequence identities >35% (with at least 60% of the larger domain equivalent to the smaller), indicating highly similar structures and functions.

CATH is accessible for keyword interrogation via UCL's Biomolecular Structure and Modelling Unit Web server.

65

Summary

3.7.3 PDBsum

A major resource for accessing structural information is PDBsum (Laskowski *et al.*, 1997), a Web-based compendium maintained at UCL. PDBsum provides summaries and analyses of all structures in the PDB. Each summary gives an at-a-glance overview of the contents of a PDB entry in terms of **resolution** and **R-factor**, numbers of protein chains, ligands, metal ions, secondary structure, fold cartoons and ligand interactions, etc. This is vital, not only for visualising the structures concealed in PDB files, but also for drawing together in a single resource information at the 1D (sequence), 2D (motif) and 3D (structure) levels. Resources of this type will become more and more important as visualisation techniques improve, and new-generation software allows more direct interaction with their contents.

PDBsum is accessible for keyword interrogation via UCL's Biomolecular Structure and Modelling Unit Web server.

Advances in computer technology will play a major role in simplifying the task of sequence analysis in the future; developments such as CORBA, which facilitates distributed programming, and the Internet object-orientated programming language Java are poised to create a new generation of interactive tools that, for the first time, allow seamless integration of remote information systems at the desktop. Software that provides both 'intelligent' consensus views of the results and access to the raw search data, will cater, at the same time, for the less experienced and for the expert user. In addition, interactive 1D, 2D and 3D visualisation tools will offer new ways of interacting with dry computer outputs, helping to transform sequence, motif and structure information into biological knowledge.

3.8 Summary

- Databases are used to store the vast amounts of information issuing from the genome projects. There are many different types of database, but for routine protein sequence analysis, primary, secondary and composite databases are initially the most important.

- Primary databases contain sequence data (nucleic acid or protein). SWISS-PROT is the most highly annotated protein sequence database.

- Composite databases amalgamate a variety of different primary sources and are hence efficient to search because they obviate the need to interrogate multiple resources.

- Different composite databases use different primary sources and different redundancy criteria in their amalgamation procedures.

- Secondary databases contain pattern data, i.e., diagnostic signatures for protein families. These signatures encode the most highly conserved features of multiply aligned sequences, which are often crucial to the structure or function of the protein.

- Different sequence analysis methods have given rise to different pattern databases: the main approaches exploit single motifs (e.g., regular expressions), multiple motifs (e.g., fingerprints) and full domain alignments (e.g., Hidden Markov Models).

- There is some overlap in content between the secondary databases, but even their combined total is well short of the estimated 1 000–10 000 protein families. Pattern database growth is slow because the addition of detailed family annotation is very time consuming. PROSITE and PRINTS are the only comprehensively, manually annotated secondary databases.

- To address the annotation bottleneck, the secondary database curators are together creating a unified database of protein families known as InterPro. It is hoped that InterPro will make sequence analysis more straightforward in the future.

- Protein structure-related resources may provide a useful adjunct to the databases available for sequence analysis, and will become increasingly important as more 3D structures are solved.

3.9 Further reading

Nomenclature
ATTWOOD, T.K. (1997) Exploring the language of bioinformatics. In *Oxford Dictionary of Biochemistry and Molecular Biology,* Stanbury, H. (ed.), pp. 715–723. Oxford University Press.

Secondary pattern databases
ATTWOOD, T.K., BECK, M.E., FLOWER, D.R., SCORDIS, P. and SELLEY, J. (1998) The PRINTS protein fingerprint database in its fifth year. *Nucleic Acids Research,* **26**(1), 304–308.

BAIROCH, A., BUCHER, P. and HOFMANN, K. (1997) The PROSITE database, its status in 1997. *Nucleic Acids Research,* **25**(1), 217–221.

HENIKOFF, S., PIETROKOVSKI, S. and HENIKOFF, J.G. (1998) Superior performance in protein homology detection with the Blocks Database servers. *Nucleic Acids Research,* **26**(1), 309–312.

NEVILL-MANNING, C.G., WU, T.D. and BRUTLAG, D.L. (1998) Highly specific protein sequence motifs for genome analysis. *Proceedings of the National Academy of Sciences, USA,* **95**, 5865-5871.

SONNHAMMER, E.L.L., EDDY, S.R., BIRNEY, E., BATEMAN, A. and DURBIN, R. (1998) Pfam: multiple sequence alignments and HMM-profiles of protein domains. *Nucleic Acids Research,* **26**(1), 320–322.

Primary sequence databases

BAIROCH, A. and APWEILER, R. (1998) The SWISS-PROT protein sequence data bank and its supplement TrEMBL in 1998. *Nucleic Acids Research,* **26**(1), 38–42.

BARKER, W.C., GARAVELLI, J.S., HAFT, D.H., HUNT, L.T., MARZEC, C.R., ORCUTT, B.C., SRINIVASARAO, G.Y., YEH, L.-S.L., LEDLEY, R.S., MEWES, H.W., PFEIFFER, F. and TSUGITA, A. (1998) The PIR-International Protein Sequence Database. *Nucleic Acids Research,* **26**(1), 27–32.

MEWES, H.W., HANI, J., PFEIFFER, F. and FRISHMAN, D. (1998) MIPS: a database for protein sequences and complete genomes. *Nucleic Acids Research,* **26**(1), 33–37.

NAMBOODIRI, K., PATTABIRAMAN, N., LOWREY, A., GABER, B., GEORGE, D.G. and BARKER, W.C. (1990) NRL-3D – A sequence-structure database. *Biophysical Journal,* **57**(2), p. A406.

Composite sequence databases

BLEASBY, A.J., AKRIGG, D. and ATTWOOD, T.K. (1994) OWL – A non-redundant, composite protein sequence database. *Nucleic Acids Research,* **22**(17), 3574–3577.

Information retrieval tools

ETZOLD, T., ULYANOV, A. and ARGOS, P. (1996) SRS – Information-retrieval system for molecular-biology data-banks. *Methods in Enzymology,* **266**, 114–128.

Web servers

HENIKOFF, S., ENDOW, S.A. and GREENE, E.A. (1996) Connecting protein family resources using the proWeb network. *Trends in Biochemical Science,* **21**, 444–445.

Structure classification resources

LASKOWSKI, R.A., HUTCHINSON, E.G., MICHIE, A.D., WALLACE, A.C., JONES, M.L. and THORNTON, J.M. (1997) PDBsum: a Web-based database of summaries and analyses of all PDB structures. *Trends in Biochemical Science,* **22**, 488–490.

MURZIN, A.G., BRENNER, S.E., HUBBARD, T. and CHOTHIA, C. (1995) scop: a structural classification of proteins database for the investigation of sequences and structures. *Journal of Molecular Biology,* **247**, 536–540.

ORENGO, C.A., MICHIE, A.D., JONES, S., JONES, D.T., SWINDELLS, M.B. and THORNTON, J.M. (1997) CATH – a hierarchic classification of protein domain structures. *Structure,* **5**(8), 1093–1108.

3.10   Web addresses

GenBank	http://www.ncbi.nlm.nih.gov/Web/Genbank/
EMBL	http://www.ebi.ac.uk/ebi_docs/embl_db/ebi/topembl.html
DDBJ	http://www.ddbj.nig.ac.jp/
PIR	http://nbrfa.georgetown.edu/pir/
MIPS	http://www.mips.biochem.mpg.de/
SWISS-PROT	http://expasy.hcuge.ch/sprot/sprot-top.html
NRL-3D	http://www.nbrfa.georgetown.edu/pir/nrl3d.html
OWL	http://www.biochem.ucl.ac.uk/bsm/dbbrowser/OWL/
PROSITE	http://expasy.hcuge.ch/sprot/prosite.html
PRINTS	http://www.biochem.ucl.ac.uk/bsm/dbbrowser/PRINTS/
BLOCKS	http://www.blocks.fhcrc.org/
Profiles	http://ulrec3.unil.ch/software/PFSCAN_form.html
Pfam	http://www.sanger.ac.uk/Software/Pfam/
IDENTIFY	http://dna.Stanford.EDU/identify/
proWeb	http://www.proweb.org/kinesin/ProWeb.html
SCOP	http://scop.mrc-lmb.cam.ac.uk/scop/
CATH	http://www.biochem.ucl.ac.uk/bsm/cath/
PDBsum	http://www.biochem.ucl.ac.uk/bsm/pdbsum/

CHAPTER FOUR

Genome information resources

4.1 Introduction

In Chapter 3, we discussed several different types of biological database, focusing on protein information resources and specifically on the structures of particular database formats (e.g., taking SWISS-PROT as a model). Here, we take a closer look at DNA sequence data repositories, including the primary producers (GenBank, EMBL, DDBJ) and a range of specialist genome information resources. As the format of EMBL entries is consistent with that already described for SWISS-PROT, we examine instead the structure of GenBank.

4.2 DNA sequence databases

4.2.1 EMBL

EMBL, the nucleotide sequence database from the European Bioinformatics Institute (EBI), includes sequences both from direct author submissions and genome sequencing groups, and from the scientific literature and patent applications (Stoesser *et al.*, 1998). The database is produced in collaboration with DDBJ and GenBank; the participating groups each collect a portion of the total sequence data reported worldwide, and all new and updated entries are then exchanged between the groups on a daily basis.

The rate of growth of DNA databases has been following an exponential trend, with a doubling time now estimated to be ~9–12 months. In January 1998, EMBL contained more than a million entries, representing more than 15 500 species, but with model systems predominating (*Homo sapiens*, *Caenorhabditis elegans*, *Saccharomyces cerevisiae*, *Mus musculus* and *Arabidopsis thaliana* together constitute more than 50% of the resource).

Information can be retrieved from EMBL using the SRS Sequence Retrieval System (Etzold *et al.*, 1996); this links the principal DNA and protein sequence databases with motif, structure, mapping and other specialist databases, and includes links to the MEDLINE facility. EMBL may be searched with query sequences via the EBI's Web interfaces to the BLAST and FastA programs.

4.2.2 DDBJ

DDBJ is the DNA Data Bank of Japan, which began in 1986 as a collaboration with EMBL and GenBank (Tateno *et al.*, 1998). The database is produced, maintained and distributed at the National Institute of Genetics; sequences may be submitted to it from all corners of the world by means of a Web-based data-submission tool. The Web is also used to provide standard search tools, such as FastA and BLAST.

4.2.3 GenBank

GenBank, the DNA database from the National Center for Biotechnology Information (NCBI), incorporates sequences from publicly available sources, primarily from direct author submissions and large-scale sequencing projects (Benson *et al.*, 1998). To help ensure comprehensive coverage, the resource exchanges data with both the EMBL Data Library and DDBJ.

The increasing size of the database (Box 4.1), coupled with the diversity of data sources available, have made it convenient to split GenBank into smaller, discrete divisions (17 to date); these are summarised in Table 4.1.

Table 4.1 The three-letter codes for each of the 17 divisions of GenBank.

Division	Sequence subset
PRI	Primate
ROD	Rodent
MAM	Other mammalian
VRT	Other vertebrate
INV	Invertebrate
PLN	Plant, fungal, algal
BCT	Bacterial
RNA	Structural RNA
VRL	Viral
PHG	Bacteriophage
SYN	Synthetic
UNA	Unannotated
EST	EST (Expressed Sequence Tags)
PAT	Patent
STS	STS (Sequence Tagged Sites)
GSS	GSS (Genome Survey Sequences)
HTG	HTG (High Throughput Genomic Sequences)

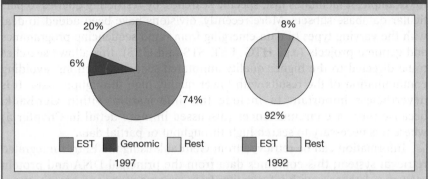

1997 — EST 20%, Genomic 6%, Rest 74%

1992 — EST 8%, Rest 92%

The pie charts above illustrate the sequence entry statistics for GenBank releases 104 (December 1997) and 74 (December 1992). The data have been divided into the following categories: partial sequences, including EST and **Sequence Tag Site (STS)** sections; genomic sequences, including HTG and GSS sections; and all the rest of the sequences. It is clear from a comparison of the charts that this five-year period has witnessed a near 10-fold increase in the quantity of sequence tag data contributing to the database. In addition, new sections have been provided to house the sequence information emerging from various genome projects world-wide. Viewed in this manner, it is evident that the number of partial sequence entries now significantly outweighs the number of other entries.

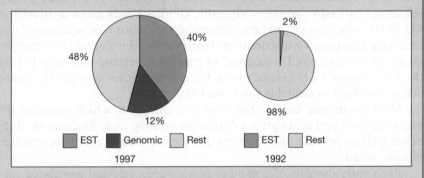

1997 — EST 40%, Genomic 12%, Rest 48%

1992 — EST 2%, Rest 98%

In these charts, base count statistics from the same releases have been compared, and data have been divided into the categories described above. Clearly, in 1992, ESTs represented an insignificant portion of the entire database, but by 1997 this proportion had grown to 40%; this means that in the same five-year period, the quantity of sequence tag data has shown a 20-fold increase. It is evident that, although partial sequences now contribute a sizeable proportion of the nucleotides, more bases are in fact derived from the genomic and other sections of the database. This is because ESTs (and other partial data, such as STSs) are usually short sequences (<400 bases), while full-length sequences, deposited in the sections making up the rest of the database, tend to be much longer (on average several **kilobases**). Thus, base counts give a fairer picture of the information content of the database; while it might appear that sequence tags dominate the entry statistics, the overall sequence information they contribute is still less than that of the higher quality sequences in the other major divisions (PRI, MAM, PLN, etc.).

This somewhat artificial separation may be useful for a number of reasons: for example, it facilitates fast, specific searches, by restricting queries to particular database subsets. More recently, divisions have been added to deal with the varying types of data emerging from rapid sequencing programmes and genome projects (e.g., HTG, EST, STS and GSS); this allows searches to be directed to the higher quality annotated sequence sections, avoiding contamination of the results with lower quality high-throughput data. It is nevertheless important to include this information within GenBank, because there are circumstances (discussed in more detail in Chapter 5) where it is necessary to search high-throughput or partial data.

Information can be retrieved from GenBank using the Entrez integrated retrieval system; this combines data from the principal DNA and protein sequence databases with information from genome maps and protein structures. Additional information on the sequences can be accessed via the MEDLINE facility, which provides abstracts from the original published articles. GenBank may be searched with user query sequences by means of the NCBI's Web interface to the BLAST suite of programs (BLAST is described in further detail in Chapter 6).

The structure of GenBank entries

A GenBank release includes the sequence files, indices created on various database fields (e.g., author index, reference index, etc.) and information derived from the database (e.g., GenPept, a database of translated coding sequences in FastA format). Originally, GenBank was made available on CD-ROM, which proved to be a convenient mechanism for widespread and relatively inexpensive distribution. However, as the size of the database grew, the number of CDs required to contain it became unwieldy (12 for the last available CD release) both for the producers and for the users. Today, GenBank is available solely via FTP.

Most commonly used is the sequence entry file, which contains the sequence itself and descriptive information relating to it. Since many Web-based systems provide links back to this file, we now examine its structure in some detail.

Each entry consists of a number of keywords, relevant associated sub-keywords, and an optional Feature Table; its end is indicated by a // terminator. The positioning of these elements on any given line is important: keywords begin in column 1; sub-keywords begin in column 3; a code defining part of the Feature Table begins in column 5. Any line beginning with a blank character is considered a continuation from the keyword or sub-keyword above.

In Figure 4.1, keywords include LOCUS, DEFINITION, ACCESSION, NID, KEYWORDS, SOURCE, REFERENCE, FEATURES, BASE COUNT and ORIGIN. The LOCUS keyword introduces a short label for the entry that may suggest the function of the sequence (here, to the trained eye, HUMCY-CLOX suggests a human cyclooxygenase); this line summarises other relevant facts, including the number of bases, source of sequence data (mRNA), section of database (PRI) and date of submission. The DEFINITION line contains a

```
LOCUS       HUMCYCLOX   3387 bp  mRNA  PRI  31-DEC-1994
DEFINITION  Homo sapiens cyclooxygenase-2 (Cox-2) mRNA, complete cds.
ACCESSION   M90100
NID         g181253
KEYWORDS    cyclooxygenase-2; prostaglandin synthase.
SOURCE      Homo sapiens umbilical vein cDNA to mRNA.
  ORGANISM  Homo sapiens
            Eukaryotae; mitochondrial eukaryotes; Metazoa; Chordata;
            Vertebrata; Eutheria; Primates; Catarrhini; Hominidae; Homo.
REFERENCE   1 (bases 1 to 3387)
  AUTHORS   Hla,T. and Neilson,K.
  TITLE     Human cyclooxygenase-2 cDNA
  JOURNAL   Proc. Natl. Acad. Sci. U.S.A. 89 (16), 7384-7388 (1992)
  MEDLINE   92366465
FEATURES             Location/Qualifiers
     source          1..3387
                     /organism="Homo sapiens"
                     /db_xref="taxon:9606"
                     /cell_type="endothelial"
                     /tissue_type="umbilical vein"
     5'UTR           1..97
                     /gene="Cox-2"
     gene            1..3387
                     /gene="Cox-2"
     CDS             98..1912
                     /gene="Cox-2"
                     /EC_number="1.14.99.1"
                     /codon_start=1
                     /product="cyclooxygenase-2"
                     /db_xref="PID:g181254"
                     /translation="MLARALLLCAVLALSHTANPCCSHPCQNRGVCMSVGFDQYKCDC
                     TRTGFYGENCSTPEFLTRIKLFLKPTPNTVHYILTHFKGFWNVVNNIPFLRNAIMSYV
                     …
                     EYRKRFMLKPYESFEELTGEKEMSAELEALYGDIDAVELYPALLVEKPRPDAIFGETM
                     VEVGAPFSLKGLMGNVICSPAYWKPSTFGGEVGFQIINTASIQSLICNNVKGCPFTSF
                     SVPDPELIKTVTINASSSRSGLDDINPTVLLKERSTEL"
     sig_peptide     98..148
                     /gene="Cox-2"
     mat_peptide     149..1909
                     /gene="Cox-2"
                     /EC_number="1.14.99.1"
                     /product="cyclooxygenase-2"
     3'UTR           1913..3387
                     /gene="Cox-2"
     polyA_signal    3369..3374
                     /gene="Cox-2"
BASE COUNT     1010 a    712 c    633 g    1032 t
ORIGIN
     1 gtccaggaac tcctcagcag cgcctccttc agctccacag ccagacgccc tcagacagca
    61 aagcctaccc ccgcgccgcg ccctgcccgc cgctgcgatg ctcgtcccgc ccctgctgct
   121 gtgcgcggtc ctggcgctca gccatacagc aaatccttgc tgttccacc catgtcaaaa
   …
  3301 tacctgaact tttgcaagtt ttcaggtaaa cctcagctca ggactgctat ttagctcctc
  3361 ttaagaagat taaaaaaaaa aaaaaag
//
```

Figure 4.1 Example GenBank entry illustrating the use of keywords, sub-keywords and the Feature Table to express information on the structure of the cDNA for Cox-2. Both the protein translation in the Feature Table and the nucleotide sequence have been abbreviated (…) for the figure.

concise description of the sequence (in this example, *Homo sapiens* cyclooxygenase-2 (Cox-2) mRNA, complete cds).

Following this, the ACCESSION line gives the accession number, a unique, constant code assigned to each entry (here M90100). The NID line supplies a nucleotide identifier (g181253), intended to provide a unique reference to the current version of the sequence information; this allows the sequence to be revised but still be associated with the same locus name and accession number. Through a collaborative arrangement between the principal DNA database providers, a new identifier (in the form 'accession.version'), which expresses the relationship between entry and sequence version more explicitly, is being introduced; this will lead to the eventual phasing out of the NID and associated protein identifier, PID.

The KEYWORDS line introduces a list of short phrases, assigned by the author, describing gene products and other relevant information about the entry (in this example, cyclooxygenase-2; prostaglandin synthase). The SOURCE record provides information on the tissue from which the data have been derived (here, umbilical vein); the sub-keyword ORGANISM illustrates the biological classification of the source organism (here, *Homo sapiens,* Eukaryotae, etc. – see figure). REFERENCE records indicate the portion of sequence data to which the cited literature refers; sub-keywords AUTHORS, TITLE and JOURNAL provide a structure for the citation; the MEDLINE sub-keyword is a pointer to an online medical literature information resource, which allows the abstract of a given article to be viewed.

The FEATURES keyword introduces a section that has its own structure, and whose purpose is to describe properties of the sequence in detail – this is the Feature Table. Within the table, links to other databases are made through the '/db_xref' qualifier (here, for example, we see links to a taxonomic database (taxon:9606) and to a protein sequence database (PID:g181254)); co-ordinates are given for the 5'-untranslated region (1–97), for the coding sequence (98–1912), for the 3'-untranslated region (1913–3387), the polyadenylation sequence (3369–3374), etc.; and the protein translation and location of the signal and mature peptides are also given. This example is not exhaustive, but serves to indicate the level of detail that can be represented in the Feature Table.

The entry continues with the BASE COUNT record, which details the frequency of occurrence of the different base types in the sequence (here, 1010 A, 712 C, 633 G and 1032 T). The ORIGIN line notes, where possible, the location of the first base of the sequence within the genome. The nucleotide sequence itself follows, and the entry is terminated by the // marker.

4.2.4 dbEST

EST data are held in the dbEST database, which maintains its own format and identification number system. The sequence data, together with a summary of the dbEST annotation, are also distributed as a sub-section of the primary DNA databases. The dbEST release of 8 May 1998 contained more than 1.6 million ESTs, including more than 1 million sequences from *Homo sapiens,* and 300 000 from *Mus musculus* and *Mus domesticus.*

The Genome Sequence DataBase (GSDB) is produced by the National Center for Genome Resources at Santa Fe, New Mexico. GSDB creates, maintains and distributes a complete collection of DNA sequences and related information to meet the needs of major genome sequencing laboratories. The resource operates as an online, client–server relational database, offering facilities for large-scale producers to submit sequence data to it remotely. Data acquired in this manner are subjected to quality control checks prior to distribution.

The format of GSDB entries is consistent with that already described in detail for GenBank (see Section 4.2.3 for keyword definitions). The principal difference between the two formats is the inclusion of the GSDBID keyword, which implements the tracking mechanism for GSDB accessions, as illustrated in Figure 4.2.

The database is accessible either via the Web, or using relational database client–server facilities; in both cases, familiarity with the database access language, SQL (Structured Query Language), is useful.

4.3 Specialised genomic resources

In addition to the comprehensive DNA sequence databases, which cover aspects of genomic data, from complete chromosomes to individual gene products, there exist a variety of more specialised genomic resources (otherwise known as boutique databases). These tend to be linked, to some extent, with the primary DNA databases from which they may derive their data, and into which their results are usually fed. The purpose of these specialised resources is to bring a focus (a) to species-specific genomics, and (b) to particular sequencing techniques. Sequence information *per se* may not be the primary thrust of such databases – often the goal is to present a more integrated view of a particular biological system (for example, the model organisms *Saccharomyces cerevisiae, Caenorhabditis elegans, Drosophila melanogaster, Arabidopsis thaliana, Helicobacter pylori*, etc.), in which sequence data represent just one level of abstraction, and higher levels lead to an overall understanding of the genome organisation.

It is difficult to overemphasise the impact the Internet has had on the ability of scientists to present and disseminate research results from the genomic sciences. The following small selection of databases is indicative of the scope of resources currently available, from eclectic Web sites to downloadable flat-files.

4.3.1 SGD

The *Saccharomyces* Genome Database (SGD) (Cherry *et al.*, 1998) is an online resource that acts as a clearing house, bringing together information on the molecular biology and genetics of *S. cerevisiae* (commonly known as baker's, or budding, yeast). It provides Internet access to the complete

```
LOCUS       R96180 355 bp mRNA EST 11-SEP-1995
DEFINITION  yt84f11.r1 Homo sapiens cDNA clone 231021 5' similar to gb:M59979
            PROSTAGLANDIN G/H SYNTHASE 1 PRECURSOR (HUMAN);.
GSDBID      GSDB:S:319963
ACCESSION   R96180
NID         g981840
KEYWORDS    EST.
SOURCE      Homo sapiens (clone: 231021).
  ORGANISM  Homo sapiens
            Eukaryotae; Metazoa; Eumetazoa; Bilateria; Coelomata;
            Deuterostomia; Chordata; Vertebrata; Gnathostomata; Osteichthyes;
            Sarcopterygii; Choanata; Tetrapoda; Amniota; Mammalia; Theria;
            Eutheria; Archonta; Primates; Catarrhini; Hominidae; Homo.
REFERENCE   1 (bases 1 to 355)
  AUTHORS   Hillier,L., Clark,N., Dubuque,T., Elliston,K., Hawkins,M.,
            Holman,M., Hultman,M., Kucaba,T., Le,M., Lennon,G., Marra,M.,
            Parsons,J., Rifkin,L., Rohlfing,T., Soares,M., Tan,F.,
            Trevaskis,E., Waterston,R., Williamson,A., Wohldmann,P., Wilson,R.
    TITLE   The WashU-Merck EST Project
  JOURNAL   Unpublished (1995)
COMMENT

            Contact: Wilson RK
            WashU-Merck EST Project
            Washington University School of Medicine
            4444 Forest Park Parkway, Box 8501, St. Louis, MO 63108
            Tel: 314 286 1800
            Fax: 314 286 1810
            Email: est@watson.wustl.edu
            High quality sequence stops: 307
            Source: IMAGE Consortium, LLNL
            This clone is available royalty-free through LLNL ; contact the
            IMAGE Consortium (info@image.llnl.gov) for further information.

            10337-0-gland-001-1997-1.0 sequence R96180 comprises a
            single-sequence cluster by mp_d2cluster in the SANBI clustering
            of dbEST release 080496. This comment generated Fri Nov 14
            14:28:48 1997

            [Flatfile retrieved from GSDB Fri May 15 08:47:38 1998]
FEATURES            Location/Qualifiers
     mRNA           <1..>355
                    /gsdb_id="GSDB:F:404641"
                    /note="putative"
     source         1..355
                    /organism="Homo sapiens"
                    /clone="231021"
BASE COUNT      75 a    89 c    100 g    88 t    3 others
ORIGIN
        1 gttctacctg gtactgcttc ttgaaaagtg ccatccaaac tctatctttg gggagagtat
       61 gatagagatt ggggctccct tttccctcaa gggtctccta gggaatccca tctgttctcc
      121 ggagtactgg aagccgagca catttggcgg cgaggtgggc tttaacattg tcaagacggc
      181 cacactgaag aagctggtct gcctcaacac caagacctgt ccctacgttt ccttccgtgt
      241 gccggatgcc agtcaggatg atgggcctgc tgtggagcga ccatcccacg agctctgagg
      301 ggcaggaaag cagcattctg gangggagag ctttgtgctt gtcattncag antgc
//
```

Figure 4.2 Example EST sequence from dbEST, as retrieved from the GSDB server. Note that an additional keyword, GSDBID, has been introduced to provide a tracking mechanism.

S. cerevisiae genome, its genes and their products, the phenotypes of its mutants, and the literature supporting these data (this is important because this was the first, and by mid-1998 the only, eukaryotic genome to have been sequenced in its entirety). SGD aids researchers by incorporating functions to perform sequence similarity searches, utilise Web-based gene and sequence analysis resources, register a yeast gene name, display maps of genomic data, examine 3D structural data, access primer sequences that have been used successfully to clone yeast genes, and so on. The database presents information using a variety of user-friendly, dynamically created graphical displays illustrating physical, genetic and sequence feature maps.

4.3.2 UniGene

A primary goal of the Human Genome Project is to determine the complete sequence of the human genome (estimated to contain about 3 billion base pairs). However, only about 3% of the genome encodes protein, and the biological significance of the remainder is currently unknown. A transcript map is thus a vital resource in flagging those parts of the genome that are actually expressed.

UniGene attempts to provide a transcript map by utilising sets of non-redundant gene-oriented clusters derived from GenBank sequences. The collection represents genes from many organisms (compare the HGI in Section 5.9.3, which represents only human genes), each cluster relating to a unique gene and including related information, such as the tissue type in which the gene is expressed, map location, etc.

In addition to sequences of well-characterised genes, substantial numbers of novel ESTs have been included, which means that many of the sequences are partial and the corresponding genes uncharacterised. Consequently, a valuable role for the collection is in gene discovery. UniGene has also been used by experimentalists to select reagents for gene-mapping projects and large-scale **gene expression** analysis. The resource can be accessed via NCBI's home-page.

4.3.3 TDB

The TIGR database (TDB) provides a substantial suite of databases containing DNA and protein sequence, gene expression, cellular role, and protein family information, and taxonomic data for microbes, plants and humans. Specifically, the resources include a microbial database that links to worldwide and TIGR genome sequencing projects (e.g., *A. fulgidus, B. burgdorferi, H. influenzae, H. pylori, M. jannaschii* and *M. genitalium)* a parasite database (*T. brucei* and *P. falciparum*); human, mouse, rice, etc., gene index projects (see Section 5.9.3 for detailed discussion of the Human Gene Index); an *A. thaliana* database; a human genomic dataset; and so on. Some of the data are available for download via the FTP site, or may be accessed via the TIGR home-page.

4.3.4 ACeDB

ACeDB is 'A *C. elegans* DataBase' arising from the *C. elegans* genome project. The resource includes restriction maps, gene structural information (introns, exons, promoter sites, etc.), cosmid maps, sequence data, bibliographic references, and so on. The software designed to organise and browse the information, known as ACEDB, presents a graphical interface that enables the user to view genomic data at different stages of resolution, from the level of a complete chromosome down to the physical (sequence) level, as illustrated in Figure 4.3. The use of ACeDB and ACEDB to refer to both the database and the software can lead to confusion, so users should be aware of the distinction.

The software has been designed using object-oriented technology, resulting in a system with sufficient flexibility and generality to be applied easily to the analysis of data from other genome projects. It has, for example, been used to analyse data from *A. thaliana, S. cerevisiae* and various human chromosomes. In keeping with Web-based developments, a set of CGI scripts and perl modules, collectively termed webace, has been created to enable ACEDB databases to be accessed via the WWW (humace, for example, provides the Web interface to the ACEDB database of human sequence determined at the Sanger Centre).

4.4 Summary

- The principal nucleic acid sequence databases are GenBank, EMBL and DDBJ, which each collect a portion of the total sequence data reported world-wide, and exchange new and updated entries on a daily basis.

- GenBank, which is produced at the NCBI, is split into smaller, discrete divisions. This facilitates fast, specific searches by restricting queries to particular database subsets.

- During 1992–1997, the level of EST and STS data within GenBank grew 10-fold. Nevertheless, the overall sequence information contributed by such partial data was still less than that of the higher quality sequences in the other major divisions.

- In addition to the comprehensive DNA sequence databases, there is a variety of more specialised genomic resources. These so-called boutique databases bring a focus to species-specific genomics and to particular sequencing techniques.

- The scope of genomic resources available on the Internet is immense and has had an enormous impact on the ability of scientists to present and disseminate research results.

4.5 Further reading

DNA sequence databases

BENSON, D.A., BOGUSKI, M.S., LIPMAN, D.J., OSTELL, J. and OUELLETTE, B.F.F. (1998) GenBank. *Nucleic Acids Research,* **26**(1), 1–7.

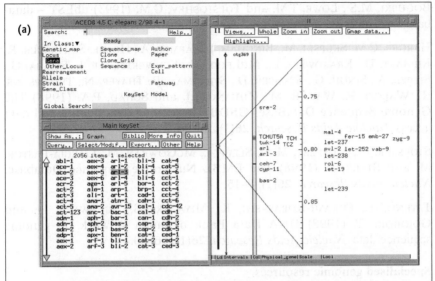

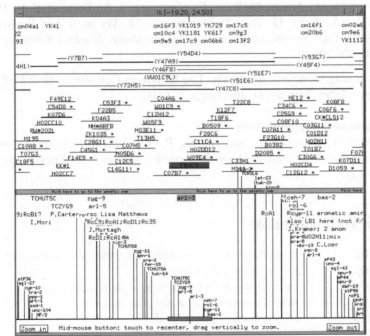

Figure 4.3 Using ACEDB to drill down through the data layers in a genome map. (a) The top left frame shows the opening window of ACEDB, in which 'Gene' has been selected. This opens the window beneath, labelled 'Main KeySet', which contains a scrolling list of gene names from which to choose (here, arl-3 was selected). The area corresponding to the locus for arl-3 is shown in the right-hand window. Clearly, there are also other genes located in the vicinity. By continuing to zoom in, at successively higher resolutions, we would end up with the sequence for the arl-3 gene. (b) High-level view of the map data in ACEDB, indicating the clones that were used to build the map (shown as overlapping horizontal lines, e.g. Y54D4).

BOGUSKI, M.S., LOWE, T.M. and TOLSTOSHEV, C.M. (1993) dbEST – database for 'expressed sequence tags'. *Nature Genetics* **4**(4), 332–333.

HARGER, C.M, SKUPSKI, M., BINGHAM, J., FARMER, A., HOISIE, S., HRABER, P., KIPHART, D., KRAKOWSKI, L., MCLEOD, M., SCHWERTFEGER, J., SELUJA, G., SIEPEL, A., SINGH, G., STAMPER, D., STEADMAN, P., THAYER, N., THOMPSONS, R., WARGO, P., WAUGH, M., ZHUANG, J.J. and SCHAD, P.A. (1998) The Genome Sequence DataBase (GSDB): improving data quality and data access. *Nucleic Acids Research*, **26**(1), 21–26.

STOESSER, G., MOSELEY, M.A., SLEEP, J., MCGOWRAN, M., GARCIA-PASTOR, M. and STERK, P. (1998) The EMBL Nucleotide Sequence Database. *Nucleic Acids Research*, **26**(1), 8–15.

TATENO, Y., FUKAMI-KOBAYASHI, K., MIYAZAKI, S., SUGAWARA, H. and GOJOBORI, T. (1998) DNA Data Bank of Japan at work on genome sequence data. *Nucleic Acids Research*, **26**(1), 16–20.

Specialised genomic resources
CHERRY, J. M., ADLER, C., BALL, C., CHERVITZ, S. A., DWIGHT, S. S., HESTER, E.T., JIA, Y., JUVIK, G., ROE, T., SCHROEDER, M., WENG, S. and BOTSTEIN, D. (1998) SGD: Saccharomyces Genome Database. *Nucleic Acids Research*, **26**(1), 73–79.

Advancing Genomic Research: The UniGene Collection. In *NCBI News*, August 1996, Benson, D. and Rapp, B. (eds).

Information retrieval tools
ETZOLD, T., ULYANOV, A. and ARGOS, P. (1996) SRS–Information retrieval system for molecular-biology data-banks. *Methods in Enzymology*, **266**, 114–128.

4.6 Web addresses

EMBL	http://www.ebi.ac.uk/ebi_docs/embl_db/ebi/topembl.html
DDBJ	http://www.ddbj.nig.ac.jp/
GenBank	http://www.ncbi.nlm.nih.gov/Web/Genbank/
dbEST	http://www/ncbi.nlm.nih.gov/dbEST/
GSDB	http://www.ncgr.org/gsdb
SGD	http://genome-www.stanford.edu/Saccharomyces/
UniGene	http://www.ncbi.nlm.nih.gov/UniGene/
TDB	http://www.tigr.org/tdb/tdb.html
AceDB	http://www.sanger.ac.uk/Software/Acedb/
webace	http://webace.sanger.ac.uk/

CHAPTER FIVE

DNA sequence analysis

5.1 Introduction

This chapter presents the specific motivations for analysing DNA sequences, as opposed to protein sequences. Terms in common use are defined, and the genetic code is reviewed. The concept of the hierarchy of genomic information is introduced, and the transcribed genome is examined in some detail, leading to a discussion of the Expressed Sequence Tag (EST) as a unit of sequence data, derived from rapid sequencing of cDNA libraries. Gene discovery is presented in the context of drug target discovery. Issues associated with manipulating this type of sequence information, together with possible practical solutions, are presented. Finally, three examples of producers of EST databases are profiled.

5.2 Why analyse DNA?

The most sensitive comparisons between sequences are made at the protein level; detection of distantly related sequences is easier in protein **translation**, because the redundancy of the **genetic code** of 64 codons (see Table 5.1) is reduced to 20 distinct amino acids, the functional building blocks of proteins. However, the loss of degeneracy at this level is accompanied by a loss of information that relates more directly to the evolutionary process, because proteins are a functional abstraction of genetic events that occur in DNA. This is illustrated in the analysis of silent mutations, as shown in Box 5.1.

Twenty years ago, the primary means of determining the order of amino acids in a polypeptide chain was via the low-throughput technique of chemical protein sequencing. This remains a powerful approach, for example in confirming the sequence of an expressed protein from a genetically

Table 5.1 The genetic code.

	T		C		A		G		
T	TTT	Phe	TCT	Ser	TAT	Try	TGT	Cys	T
	TTC		TCC		TAC		TGC		C
	TTA	Leu	TCA		TAA	Stop	TGA	Stop	A
	TTG		TCG		TAG		TGG	Trp	G
C	CTT	Leu	CCT	Pro	CAT	His	CGT	Arg	T
	CTC		CCC		CAC		CGC		C
	CTA		CCA		CAA	Gln	CGA		A
	CTG		CCG		CAG		CGG		G
A	ATT	Ile	ACT	Thr	AAT	Asn	AGT	Ser	T
	ATC		ACC		AAC		AGC		C
	ATA		ACA		AAA	Lys	AGA	Arg	A
	ATG	Met	ACG		AAG		AGG		G
G	GTT	Val	GCT	Ala	GAT	Asp	GGT	Gly	T
	GTC		GCC		GAC		GGC		C
	GTA		GCA		GAA	Glu	GGA		A
	GTG		GCG		GAG		GGG		G

engineered construct. However, the lack of scalability of the method is a rate-determining step in generating large quantities of data. More recently, the advent of high-throughput automated fluorescent DNA sequencing technology has led to the rapid accumulation of sequence information, and provides the basis for abundant computationally derived protein sequence data. Analysis of DNA sequences underpins a number of aspects of research: these include, for example, detection of phylogenetic relationships; genetic engineering using restriction site mapping; determination of gene structure through **intron/exon** prediction; inference of protein coding sequence through **open reading frame** (ORF) analysis; etc.. Some of these techniques are outlined in the following sections.

5.3 Gene structure and DNA sequences

There are certain key features of eukaryotic genes (see Box 5.2) that need to be assimilated in order to understand the impact of gene structure on sequence analysis: these include, for example, introns, exons, coding sequences (CDS), untranslated regions, etc. (**prokaryotic** genes generally lack introns and, consequently, have a somewhat simpler structure). DNA sequence databases typically contain genomic sequence data, which includes information at the level of the untranslated sequence, introns and exons,

BOX 5.1: FAMILY ANALYSIS AT THE DNA LEVEL

Phylogenetic analysis is the technique of methodically demonstrating a family relationship between species. This may be necessary, for example, to determine whether two genes are related through a process of divergent evolution from a common ancestor, or are the result of convergent evolution. This type of analysis is carried out on small sections of aligned DNA taken from the same gene in the various species under consideration, rather than on protein sequences derived from them. DNA is used because the pattern of mutations, insertions and deletions at the nucleotide level is definitive. **Silent mutations,** i.e. mutations at the DNA level that do not result in an amino acid substitution at the protein level (because of the redundancy of the genetic code), are automatically incorporated into the analysis; these would not be detectable in a purely protein-level treatment.

Phylogenetic relationships are often represented graphically, either in the form of phylogenetic trees (often unrooted), in which evolutionary distance is measured in terms of horizontal branch length, or as dendrograms, in which evolutionary distance is measured along the length of line segments, as shown; in spite of their different appearances, the tree and the dendrogram depict the relationship between the same five sequences.

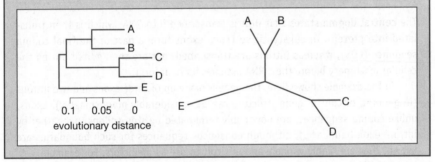

Gene structure and DNA sequences

mRNA, cDNA and translations. Thus it can be seen that so-called DNA databases in fact contain a variety of different types of data that cannot all be treated alike. These subtleties affect the manner in which search results must be interpreted (for example, searching cDNA for intron–exon boundaries, which only occur in genomic DNA, would be a rather futile exercise).

5.3.1 Untranslated regions

Untranslated regions (UTRs) occur in both DNA and RNA; they are portions of the sequence flanking the CDS that are not translated into protein (see Box 5.2). Untranslated sequence, particularly at the 3' end, is highly specific both to the gene and to the species from which the sequence is derived.

5.3.2 Conceptual translation

Given a piece of DNA sequence, and knowing the genetic code, it is possible to translate the DNA into protein by looking up successive codons in a

BOX 5.2: THE CENTRAL DOGMA OF MOLECULAR BIOLOGY

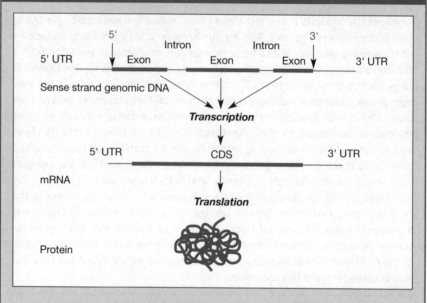

The **central dogma** states that DNA is transcribed into RNA, which is then translated into protein. In eukaryotic systems, exons form a part of the final coding sequence (CDS), whereas introns are transcribed, but are then edited out by the cellular machinery before the mRNA assumes its final form.

In the example shown here, the gene is made up of three exons and two introns (in general, of course, gene structure can be considerably more complex). Exons, unlike coding sequences, are not simply terminated by stop codons, but rather by intron–exon boundaries; although consensus sequences for such boundaries are known, they are highly variable and difficult to use as predictors. For example, a 5' intron junction may have the sequence 'AGGTAAGT', while a 3' boundary may have the sequence PyPyPyPyPyPyNCAG (where Py = pyrimidine and N = any base).

The untranslated regions (UTRs) occur at either end of the gene; if **transcription** begins at the 5' end of the sequence, then the 5' UTR contains **promoter** sites (such as the TATA box), and the 3' UTR follows the stop codon. It is important to note that promoters occur in the UTR immediately **upstream** from the site of the initiation of translation. 'Upstream' is relative to the direction of translation and should be interpreted as being further back from the point of initiation.

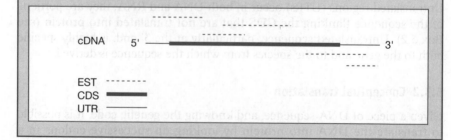

BOX 5.2: CONTINUED

In the mRNA, the start codon may be flanked by a Kozak sequence (CCGCCAUGG), which gives additional confidence to the prediction of the start of the CDS.

If a **library** is constructed, **complementary DNA (cDNA)** is run off from the mRNA stage, using reverse transcriptase. ESTs are then generated using a single read of each clone on an automated sequencing system. Some protocols use **primers** that generate two reads per clone, one from the 5' region and the other from the 3' region. This process permits clustering of related ESTs using sequence similarity in the 3' region, where 5' sequences do not overlap, and there appear to be no matches with the CDS of a full-length sequence.

genetic code table (see Table 5.1); this is termed the conceptual translation. It is important to distinguish between sequences for which the translation has some biochemical support, and those that are simply derived theoretically or computationally; the use of the term 'conceptual' indicates that the specified translation is only theoretical and carries no experimental validation.

In an arbitrary DNA sequence, it is not known whether the first base marks the start of the CDS, so it is always essential to carry out a **six-frame translation.** There are three forward frames, which are achieved by beginning to translate at the first, second and third bases respectively; the three reverse frames are determined by reversing the DNA sequence and again beginning on the first, second and third bases. Thus, for any piece of DNA, the result of a six-frame translation is six potential protein sequences, as shown in Figure 5.1.

5.4 Features of DNA sequence analysis

5.4.1 Detecting open reading frames

The question now arises, 'Which is the correct reading frame?' This is normally deemed to be the longest frame uninterrupted by a stop codon (TGA, TAA or TAG – see Table 5.1). Such a frame is known as an open reading frame, or ORF. Finding the end of an ORF is easier than finding its beginning. Usually, the initial codon in the CDS is that for methionine (ATG); but methionine is also a common residue within the CDS, so its presence is not an absolute indicator of ORF initiation. Consequently, it is usually necessary to use additional techniques to detect where the 5' untranslated sequence ends.

Several features may be used as indicators of potential protein coding regions in DNA. As already mentioned, one of these is sufficient ORF length (based on the premise that long ORFs rarely occur by chance). Recognition of flanking Kozak sequences may also be helpful in pinpointing the start of the CDS (see Box 5.2). In addition, patterns of codon usage differ in coding and non-coding regions. Specifically, the use of codons for particular amino acids varies according to species, and codon-use rules break down in regions of sequence that are not destined to be translated. Thus, codon-usage statistics can be used to infer both 5' and 3' untranslated

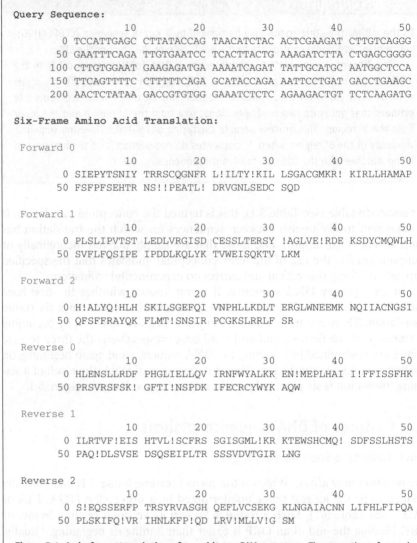

```
Query Sequence:
                10        20        30        40        50
    0 TCCATTGAGC CTTATACCAG TAACATCTAC ACTCGAAGAT CTTGTCAGGG
   50 GAATTTCAGA TTGTGAATCC TCACTTACTG AAAGATCTTA CTGAGCGGGG
  100 CTTGTGGAAT GAAGAGATGA AAAATCAGAT TATTGCATGC AATGGCTCCA
  150 TTCAGTTTTC CTTTTTCAGA GCATACCAGA AATTCCTGAT GACCTGAAGC
  200 AACTCTATAA GACCGTGTGG GAAATCTCTC AGAAGACTGT TCTCAAGATG

Six-Frame Amino Acid Translation:

Forward 0
                10        20        30        40        50
    0 SIEPYTSNIY TRRSCQGNFR L!ILTY!KIL LSGACGMKR! KIRLLHAMAP
   50 FSFPFSEHTR NS!!PEATL! DRVGNLSEDC SQD

Forward 1
                10        20        30        40        50
    0 PLSLIPVTST LEDLVRGISD CESSLTERSY !AGLVE!RDE KSDYCMQWLH
   50 SVFLFQSIPE IPDDLKQLYK TVWEISQKTV LKM

Forward 2
                10        20        30        40        50
    0 H!ALYQ!HLH SKILSGEFQI VNPHLLKDLT ERGLWNEEMK NQIIACNGSI
   50 QFSFFRAYQK FLMT!SNSIR PCGKSLRRLF SR

Reverse 0
                10        20        30        40        50
    0 HLENSLLRDF PHGLIELLQV IRNFWYALKK EN!MEPLHAI I!FFISSFHK
   50 PRSVRSFSK! GFTI!NSPDK IFECRCYWYK AQW

Reverse 1
                10        20        30        40        50
    0 ILRTVF!EIS HTVL!SCFRS SGISGML!KR KTEWSHCMQ! SDFSSLHSTS
   50 PAQ!DLSVSE DSQSEIPLTR SSSVDVTGIR LNG

Reverse 2
                10        20        30        40        50
    0 S!EQSSERFP TRSYRVASGH QEFLVCSEKG KLNGAIACNN LIFHLFIPQA
   50 PLSKIFQ!VR IHNLKFP!QD LRV!MLLV!G SM
```

Figure 5.1 A six-frame translation of an arbitrary DNA sequence. There are three forward and three reverse translations, leading to six possible protein sequences, only one of which is likely to be real. The challenge is to discover which of the translation products is correct. '!' denotes a stop codon.

regions, and to assist the detection of mistranslations, because there is an uncharacteristically high representation of rarely used codons in these regions. Table 5.2 illustrates the considerable variability in selection of codons that different organisms employ for a particular amino acid.

In addition to their characteristic pattern of codon usage, many organisms show a general preference for G or C over A or T in the third base (wobble) position of a codon. The consequent bias towards G/C in this base can further contribute to diagnosis of ORFs.

Table 5.2 Percentage use of codons for serine in a variety of model organisms. As shown, there are six possible codons for serine, which in principle could be used with equal frequency whenever serine is specified in a CDS. In practice, however, organisms are highly selective in the particular codons they use. The characteristic differences in usage reflected here can be used to help diagnose regions of DNA that may code for protein.

Codon	E. coli	D. melanogaster	H. sapiens	Z. mays	S. cerevisiae
AGT	3	1	10	3	5
AGC	20	23	34	30	4
TCG	4	17	9	22	1
TCA	2	2	5	4	6
TCT	34	9	13	4	52
TCC	37	48	28	37	33

Finally, in the region upstream of the start codon of prokaryotic genes, detection of ribosome binding sites, which help to direct ribosomes to the correct translation start positions, is considered to be a powerful ORF indicator. But, ultimately, perhaps the surest way of predicting a gene is by alignment with a homologous protein sequence.

5.4.2 Understanding the effect of introns and exons

The genes of eukaryotes are characterised by regions that contribute towards the CDS, known as exons, and those that do not, known as introns (see Box 5.2). One consequence of the presence of exons and introns in eukaryotic genes is that potential **gene products** can be of different lengths, because not all exons may be represented in the final transcribed mRNA (although the order of exons that are included is preserved). When the mRNA editing process results in different translated polypeptides, the resulting proteins are known as **splice variants** or **alternatively spliced forms.** Thus, results of database searches with cDNA or mRNA (transcription-level information) that appear to indicate substantial deletions in matches to the query sequence could, in fact, be the result of alternative splicing.

5.4.3 DNA sequence assembly

One further aspect of the analysis of DNA sequences is the process of determining the nucleotide sequence of a **clone.** In an experiment to clone a specific gene whose sequence is already known (for example, where quantities of a gene product are required for use in a biochemical assay), it is necessary to check that the cloned sequence is indeed identical to the published one. If this should turn out not to be so, experiments must be designed to correct the sequence. **Cloning** errors can arise, for example, as a result of using incorrect primers for the cloning step, or the use of a low-fidelity enzyme in a **polymerase chain reaction** (PCR) experiment.

A cDNA clone is synthesised using mRNA as a template. The clone is then sequenced by designing primers to known oligonucleotides present in the **cloning vector** flanking the inserted DNA. When the primers **hybridise** to the corresponding sequences, they are extended in a chain synthesis reaction using the inserted sequence as template, as shown in Figure 5.2.

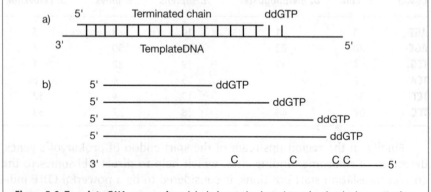

Figure 5.2 Template DNA sequencing: (a) chain synthesis and termination by incorporation of ddGTP; (b) the family of chains terminated at different positions by ddGTP. Since G pairs with C the template sequence contains C at each of these positions.

The reaction is terminated by the incorporation of a dideoxynucleotide (ddATP, ddTTP, ddGTP or ddCTP). Not all the chains terminate at the same base, however, as normal bases (dATP, dTTP, dGTP or dCTP) are also present in the reaction mixture. The result is a series of fragments for each primer, all of different lengths because they have been terminated at different base positions. The generated fragments are run on standard radioactive sequencing gels, or fluorescent sequencing machines, as appropriate, to determine the order of bases in the sequence. It is unusual to be able to sequence a complete CDS in one run, so overlapping fragments are built up in a multiple alignment, a process known as sequence assembly.

The assembler program builds a **consensus sequence** for the clone, according to a weighting given to each nucleotide position in the sequence. Parameters are set in the assembler for the number of mismatches allowed per position. Normally, a degree of redundancy in sequence coverage is required; for example, at least two base reads per position on each strand (plus and minus) give a high level of confidence in the resulting sequence, as illustrated in Figure 5.3. Conversely, part of a consensus derived from a single read on one strand (i.e., no overlapping fragments) would give only a low level of confidence in the assembled sequence.

Sequencing and degree of confidence are the result of time and patience, especially where automated fluorescent sequencing systems are employed for high throughput (see Box 5.3). Ultimately, good-quality finished sequence requires a skilled analyst, many hours of interpretation of

```
          AACCGTTTACGAAACCAGGTGC
          AACCGTTTACGAAACCAGGTGCGCGCCCGCGGGAAT
          AACCGTTTACGAACCCAGGTGC
      AACCGTTTACGAAaCCAGGTGCGCGCGCGcGGGAATCCTAAAAA
                     CGCGCGCGCGGGAATCCTAAAAA
                     TGCGCGCGCGAGGGAATCCTAAAAA
```

Figure 5.3 Example of three plus-strand and two minus-strand reads contributing to a consensus DNA sequence as a result of sequence assembly. Here, there are two positions where mismatches have resulted in lower confidence in the consensus base. Further reads are required, and/or visual inspection of the sequencing chromatogram, to resolve the ambiguities. Normally, for fully validated sequence confirmation, complete reads on the plus and minus strands are expected.

chromatogram data, and a reliable assembly program. Understanding the limitations of the sequencing protocol, effects of GC-rich regions (resulting in high secondary structure and consequently awkward reads), repetitive sequence, etc., all make sequence assembly a highly skilled pursuit.

5.5 Issues in the interpretation of EST searches

The major aspects of the analysis of full-length DNA sequences were introduced in Section 5.4. However, we do not always have the luxury of full-length information; indeed a large part of currently available DNA data is made up of partial sequences, the majority of which are Expressed Sequence Tags (ESTs). The manner in which ESTs are created is discussed in Section 5.8; here, we consider the essential properties of ESTs, and their impact on sequence interpretation. In analysing ESTs the following points should be borne in mind:

- The EST alphabet is five characters: ACGTN.

- There may be **phantom INDELs** resulting in translation **frameshifts.**

- The EST will often be a sub-sequence of any other sequence in the database.

- The EST may not represent part of the CDS of any gene.

5.5.1 The EST alphabet

The EST production process is normally highly automated and, typically, involves use of a fluorescent laser system that reads the sequencing gels. The resulting sequences are downloaded, often with little human intervention, to a computer system for further analysis. Although the gel analysis software is very sophisticated, it is sometimes unable to make a decision as to which base is present at a particular position in the sequence. When this happens, the base-calling software inserts an ambiguous base – this is

BOX 5.3: FLUORESCENT SEQUENCING CHROMATOGRAM INTERPRETATION

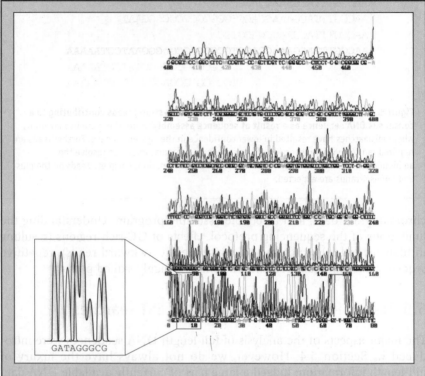

The rapid accumulation of DNA sequence data has been expedited by the introduction of fluorescent sequencing technology. Larger numbers of sequencing reactions can be carried out than has traditionally been possible with radioactive methods, and the protocols are more readily adapted to automation. When the reactions are run on a fluorescent sequencing gel, computers are used to interpret the laser-activated fluorescence and convert it into a digital form suitable for further analysis.

The result from one gel lane in a sequencing run is illustrated in the figure. Typically, 36 lanes are run on a gel at once, and higher numbers than this are constantly sought. The output consists of a series of (colour-coded) peaks, beneath which is a string of base symbols – the particular base shown is determined by the highest peak at that position of the trace. Sometimes, the software that interprets the chromatogram is unable to determine which base should be called at a specific position, so a '-' appears. Such ambiguous positions are replaced by 'N' in the resulting sequence file.

Interpretation of fluorescent chromatograms is an art. When the base-calling software has assigned an N, it is sometimes possible to call a base by eye. If a sequence is being used in an assembly process, the assembly editor normally provides the facility to view the chromatogram to increase confidence in base calling in particularly difficult areas of sequence (especially GC-rich regions, which are areas of high secondary structure content that are notoriously difficult to sequence).

IUB symbol	Represented bases
A	A
C	C
G	G
T/U	T
M	A or C
R	A or G
W	A or T
S	C or G
Y	C or T
K	G or T
V	A or C or G
H	A or C or T
D	A or G or T
B	C or G or T
X/N	G or A or T or C

normally an N, but it could be one of the other IUPAC base ambiguity symbols, as shown in Table 5.3. The consequence is a sequence in which a proportion of the symbols will be Ns.

Normal quality-control criteria in a good laboratory are expected to keep the number of Ns in a production sequence to less than 5% of the total length; the beginning and end of the sequence are trimmed to reduce ambiguity further. A typical EST will be between 200 and 500 bases in length, with modern technical advances increasing the theoretical length resulting from a single run to 1000 bases or more.

5.5.2 Insertions, deletions and frameshifts

In naïve terms, automated base-calling software simply looks for the fluorescent peaks of the four sequencing reactions in a lane on a sequencing gel. In order to increase the likelihood of finding peaks, and hence of calling bases accurately, bases are assumed to be called at regular intervals. If the physical properties of the gel, or other conditions, affect the flow across it, base calling may not actually be regular. While there is some tolerance in the software, sometimes bases are called too soon or are not called at all, resulting in phantom INDELs. At the level of subsequent DNA sequence comparisons, these have ramifications for the underlying alignment algorithms, which must consequently insert spurious INDELs into the database sequence to which the EST is being aligned. More sensitive searching techniques at the protein level are similarly hampered, the INDELs resulting in either spurious stop signals, or incorrect translation in all six frames. Such issues increase the complexity of the interpretation task.

5.5.3 Splice variants

The concept of alternatively spliced forms of a single gene was introduced earlier in the context of full-length sequence analysis. The existence of splice variants (Figure 5.4) has particular consequences for database searches with EST queries. As we have seen, alternative forms are characterised by deletions arising from the non-inclusion of all exons within a transcribed mRNA; and ESTs are noisy sequences that, as a result of sequencing errors, may not only contain ambiguous bases but also be *missing* bases. When searching a sequence database with an EST query, one of the difficulties of interpretation lies in determining whether a partially good match is the result of a sequencing error in the EST, or whether it is the reflection of a genuine genetic event, resulting from a match to an alternative form (although, in general, splice variants are usually characterised by larger deletions than are manifested by sequencing errors in ESTs).

Perhaps a more serious problem arises when an EST is sufficiently short that it falls wholly within a particular exon. In these circumstances, if alternative forms exist in the database, and both contain this exon, then there is no way of knowing which of the forms the EST represents.

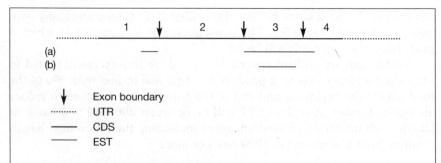

Figure 5.4 EST analysis with splice variants. An mRNA is shown, with three exon boundaries marked. Two matches are shown: (a) an EST that shows similarity to exons 1, 3 and 4, indicating that it is missing exon 2 and could thus be a splice variant; and (b) an EST that falls entirely within one exon (3), and thus we cannot be certain that it does not represent a splice variant.

5.5.4 Non-coding region ESTs

The question most often asked of an EST is 'Does this EST represent a new gene?' To answer this question, a DNA database search is usually performed. If the result shows a significant similarity to a database sequence, the normal procedure for classifying the hit will determine whether a novel gene has been found. If, however, the result shows no significant similarity, we cannot immediately assume that a new gene has been discovered; it may be that the EST represents non-coding sequence, for a known gene, that simply is not in the database.

Many mRNAs (especially human ones) have long untranslated regions at the 5' and 3' ends of the CDS. It is quite possible for an EST to be entirely from one of these non-coding regions. If we are lucky, the section of untranslated (non-coding) sequence will already be in the database. If it is, a direct match will be found, as untranslated regions are highly conserved and specific to their coding gene. If we are unlucky, no match will be found, indicating one of two possibilities: either (i) the EST represents a CDS for which there is no similar sequence in the database (still a distinct possibility), or (ii) it represents a non-coding sequence that is not in the database. It is critical to the interpretation of EST analysis that a distinction is made between these two situations.

5.6 Two approaches to gene hunting

In recent years, substantial financial resources have been expended in the search for new genes that may be linked to particular types of disease. The aim is to develop new therapies with which to combat a wide variety of prevalent disorders, such as breast cancer, asthma and neuro-degenerative diseases, to name but a few. There are two main strategies for discovering proteins that may represent suitable molecular targets, whether for small molecule drug discovery, or for gene therapy.

5.6.1 Positional cloning

The first approach for discovering disease-related genes is the technique of positional cloning. Here, the chromosome linked to the disease in question is established by analysing a population of subjects, some of whom exhibit the disease. Once a link to a chromosomal region has been established, a large part of the chromosome in the vicinity of this region (known as a **locus**) is sequenced, yielding several **megabases** of DNA. Such a locus can contain many tens of individual genes, only one of which is likely to be involved in some way in the disease process. Sequence searching and gene prediction techniques can be used to increase the efficiency of gene identification in the locus, but ultimately several genes will need to be expressed, and further experimentation (or validation) will be required to confirm which gene is actually involved in the disease. Although genes discovered in this way can be very illuminating from an academic point of view, they do not necessarily represent good drug targets (or points of therapeutic intervention). The whole process is lengthy, time-consuming and labour intensive.

5.6.2 RNA transcript analysis

The alternative approach to gene discovery, requiring much less sequencing effort and relying more heavily on the powerful search capabilities of current computer systems, examines the genes that are *actually* expressed in healthy and diseased tissue. This allows a comparison to be performed between the two states, and a process of reasoning applied to arrive at a potential drug target in a more direct way. This process analyses the mRNAs, which are used by the cellular machinery as a template for the construction of the proteins themselves.

5.6.3 The hierarchy of genomic information

The human genome is complex, consisting of about 3 billion basepairs (bp) of DNA. Yet only 3% of the DNA is coding sequence (i.e., that part of the genome that is transcribed and translated into protein). The rest of the genome consists of areas necessary for compact storage of the chromosomes, replication at cell division, the control of transcription, and so on. A large part of the work of sequence analysis is centred on analysing the products of the transcription/translation machinery of the cell, i.e., protein sequences and structures (where the latter are available). Recently, however, much industrial emphasis has been placed on the study of mRNA; this is partly because a conceptual translation into protein sequence can be generated readily, but the main reason is that mRNA molecules represent the part of the genome that is expressed in a particular cell type at a specific stage in its development. Thus, in simple terms, we have three levels of genomic information:

- the chromosomal genome (or simply the genome) – the genetic information common to every cell in the organism;

- the expressed genome (or transcriptome) – the part of the genome that is expressed in a cell at a specific stage in its development;

- the proteome – the protein molecules that interact to give the cell its individual character.

These are the three basic levels, but others could be envisaged (e.g., the metabolome). For each level, different analytical tools and interpretative skills are required. Tools appropriate for the discovery of intron/exon boundaries in genomic DNA may be interpreted in misleading ways when applied to mRNA data; sequence visualisation at the DNA, RNA and protein levels are distinct analytical techniques sharing some common threads, but again, correct interpretation of the results is paramount. The skill lies both in choosing the right tool for the job at hand, and in understanding the subtleties of the various analysis outputs.

5.7 The expression profile of a cell

Clearly, one of the goals of the genome projects is to sequence all the genes in the genome of a variety of different organisms. One way to do this is to sequence large sections of chromosomes, using either manual or automated sequencing techniques. The first chromosome from a eukaryotic organism to be sequenced in this manner was chromosome III from *Saccharomyces cerevisiae*, in an international collaboration completed in 1992. Predictions for completion of the human genome sequencing effort have been estimated to lie anywhere between 2001 and 2005, with the mouse not far behind, in 2008.

Having complete sequences and knowing what they mean are, however, two distinct stages in understanding any genome. In sequencing everything, we do not discriminate between the CDS, which we know will be transformed into a protein at some point, and all the rest of the sequence, whose precise function we can only guess at for the present.

Alternatively, we can focus on those parts of the genome that are transcribed and ultimately translated into protein. Sequences that are translated in this way make up the expressed genome for a specific cell type. Cells express a different range of genes at various stages in their development and functioning. This characteristic range of gene expression is the **expression profile** of the cell. By capturing the cell's expression profile, we can build up a picture of what levels of gene expression may be normal, or abnormal, and what the relative expression levels are between different genes within the same cell. This process also provides a rapid approach to gene discovery that complements full-blown genome sequencing projects.

5.8 cDNA libraries and ESTs

The procedure for capturing an expression profile is straightforward. First, a sample of cells is obtained (this is usually the most difficult and time-consuming step, depending on the source of cells); then RNA is extracted from the cells, and is stabilised by using reverse transcriptase to run off cDNA from the RNA template. The cDNA is transformed into a library (a **cDNA library**) suitable for use in rapid sequencing experiments. A sample of clones is selected from the library at random – e.g., 10 000 from a library with a complexity of 2 million clones. A substantial automated sequencing operation is required to produce 10 000 sequencing reactions, and then to run these on automated sequencers. The resulting data are downloaded to computers for further analysis.

The ideal result is a set of 10 000 sequences, each between 200 and 400 bases in length, representing part of the sequence of each of the 10 000 clones. In reality, some sequencing runs will fail altogether, some will fail to produce sufficient sequence data, and some will fail to produce data of appropriate quality. The sequences that emerge successfully from this process are called ESTs. The properties of ESTs and issues of interpretation are discussed in Section 5.5.

It is important to understand the statistics of library production to be confident in handling EST data. The number of clones in the library reflects the efficiency of mRNA extraction from the source of cells. Good libraries contain at least 1 million clones, and probably substantially more. Some tissues and cell types are difficult to deal with (usually the more interesting ones!) and the resulting libraries tend to be less representative. The actual number of distinct genes expressed in a cell may be a few thousand; the number varies according to cell type, with the most complex human cells expressing up to ~15 000 different genes (brain) and the simplest ~2000 (gut). Thus we have a small number of different genes ('the expressed genome') represented in a pool of 1 million clones. We then take a relatively small *single* random sample of clones for sequencing, rather than multiple samples. The relationships between these figures must be borne in mind when analysing quantitative differences between the expression levels of genes in different samples.

5.9 Different approaches to EST analysis

Various approaches to establishing libraries of ESTs for academic analysis or commercial exploitation have been developed. Here, we highlight three major sources of EST information. Much of the publicly available data are collected together into the EST sections of the EMBL Data Library and GenBank (dbEST).

Suppliers of EST information subject their data to rigorous filtering processes prior to database submission, owing to the poor quality of EST sequences. TIGR provides detailed information on its protocol, which is outlined in Section 5.9.3.

5.9.1 Merck/IMAGE

In 1994, Merck & Co. funded a research project, based at the University of Washington, to sequence 300 000 ESTs from a variety of **normalised libraries**. By choosing to use normalised libraries, quantitative information on levels of gene expression in the source tissue was sacrificed in an attempt to increase the sampling of *different* genes. The libraries chosen represent a broad cross-section of tissue types of interest to a wide variety of researchers. A representative selection of libraries sequenced is given in Table 5.4. Once identified, licensing of clones for use as reagents in further molecular biological experiments is a straightforward and cheap task. For many years Merck has sponsored the production of a drug index; this initiative has become known as the Merck Gene Index. As of May 1997, 484 421 ESTs had been submitted by the project to dbEST.

5.9.2 Incyte

Incyte Pharmaceuticals Inc. produces a database, LifeSeq, that emphasises the quantitative information derived by sequencing standard cDNA libraries. The goal here is to provide information on relative copy numbers of transcribed genes in healthy and diseased tissues, to facilitate the elucidation of potential therapeutic targets. Library sample sizes tend to be small, reflecting the desire to find the essential difference in gene expression between samples, rather than searching for every last gene by EST analysis (Incyte has other approaches to solve that problem, but discussion of them is beyond the scope of this volume). In April 1998, the size of LifeSeq was 2.5 million ESTs, representing 80 000–120 000 different genes.

Incyte's products are available on a commercial basis and tend to be licensed by large organisations. The approach taken is of interest, however,

Table 5.4 Examples of libraries sequenced as part of the Merck/WashU EST project. Soares = Bento Soares; Stratagene = Stratagene Corp.

Source of library	Tissue	Identification
Soares	placenta	Nb2HP
Soares	retina	N2b4H
Soares	breast	2NbHBst
Soares	adult brain	N2b4HB55Y
Soares	melanocyte	2NbHM
Stratagene	liver	#937224
Stratagene	lung	#937210
Stratagene	ovary	#937217
Soares	parathyroid tumour	NbHPA
Soares	senescent fibroblasts	NbHSF
Soares	ovary tumour	NbHOT
Soares	pineal gland	N3HPG
Stratagene	colon	#937204
Stratagene	corneal stroma	937222
Stratagene	pancreas	937208
Stratagene	fibroblast	937212
Stratagene	neuro-epithelium	937231
Stratagene	ovarian cancer	937219
Stratagene	colon HT-29	937221
Stratagene	endothelial cell	937223
Stratagene	skeletal muscle	937209
Stratagene	lung carcinoma	937218
Soares	germinal B-cell	NbHTGBC
Soares	testis	NHT

because this is a prime example of a 'biology *in silico*' company capitalising on genome informatics as a source of revenue.

5.9.3 TIGR

The Institute for Genomic Research (TIGR) is a not-for-profit research organisation with interests in structural, functional and comparative analysis of genomes and gene products; the range of organisms covered includes viruses, eubacteria, pathogenic bacteria, archaea and eukaryotes (both plant and animal).

TIGR Human Gene Index

An important aspect of the work at TIGR is the Human Gene Index (HGI). This Index integrates results from human gene research projects around the world, including data from dbEST and GenBank. The aim of the project is to create a non-redundant view of all human genes, and information on their expression patterns, cellular roles, functions and evolutionary relationships. The data in the HGI are freely available.

TIGR's approach has been to identify all the different human genes as rapidly as possible. To this end, it has sequenced more than 100 000 ESTs, from over 300 cDNA libraries, plus data from dbEST, and combined this information with non-redundant human transcript (HT) information, using the technique of sequence **assembly**, to generate Tentative Human Consensus (THC) sequences.

When performing an analysis on this scale, it is crucial to prepare the data thoroughly beforehand so that the risk of incorporating non-human data is minimised. For example, sequencing vectors are microbial in origin and there is a possibility that some of the vector sequence could contaminate the results, unless steps are taken to identify and remove it. In preparing data for the Index, ESTs were consequently subjected to quality-control screening to remove vector contamination, together with any poly-A, poly-T and poly-CT sequences. The minimum EST length accepted was 100 bp, with less than 3% N base calls. Additional sequences incorporated from dbEST were subjected to the same screening criteria.

The HTs were collated by removing the non-coding sequences from the *Homo sapiens* set within the primate division of GenBank. The cDNAs and CDSs were saved and redundant entries for the same gene were removed, leaving only a link to the accession number.

The TIGR assembler was used to assemble the cleaned ESTs and non-redundant HTs into **contigs.** TIGR define THCs as consensus sequences based on two or more ESTs (and possibly an HT) that overlap for at least 40 bases, with at least 95% sequence identity. THCs may contain ESTs derived from different tissues (TIGR's Expressed Gene Anatomy Database, EGAD, contains information on tissue localisation of ESTs).

In Figure 5.5, a mouse transferrin (accession number P20233) has been searched against the TIGR Gene Index. The search has been performed in translation, as the query sequence is a peptide. The output contains a graphical alignment of all the THCs that match the query sequence (in accordance with the search criteria). Notice that some THCs match in the forward orientation and others in the reverse, depending on the direction of the arrow. In the text alignment, matches are denoted by dots, and mismatches by the letter corresponding to the mismatched residue. This facilitates rapid visualisation of the extent of matches, as illustrated in the figure. Clearly, many THCs exhibit similarity with the query; some form clusters that map to different parts of the sequence; but no individual THC maps with a high degree of similarity. Note that THCs themselves represent nucleotide sequence, and it is the conceptual translation that is shown in the alignment.

The TIGR Website offers facilities for searching the index, finding out about new features added to the index, and clone ordering (via the ATCC – American Type Culture Collection). A UNIX-compressed FastA format file containing the complete, minimally redundant index is also available for transfer via anonymous FTP.

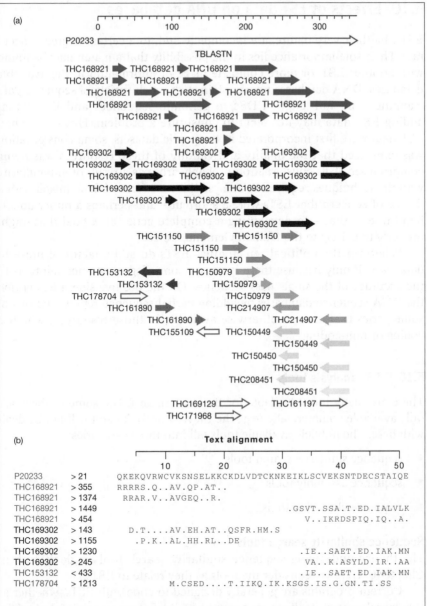

Figure 5.5 Result of searching the TIGR HGI database (March 1998 version 3.3) with part of a mouse transferrin sequence. (a) Graphical view, in which a summary of the hits in the HGI database is presented at a high level of abstraction. Arrows indicate the direction in which part of a THC has matched the query sequence, while colours aid the eye by grouping together disparate matches made by the same THC (for example, dark blue arrows at the top of the diagram all refer to THC168921 matches). The program used in the search (TBLASTN) is also indicated. (b) Text view of the same search, presented as an alignment (truncated for the figure). Dots represent identical matches with the query; non-identical residues are explicitly written out. The THC identification code is followed by a match direction indicator, > or <, and then the position of the start of the match.

5.10 Effects of EST data on DNA databases

ESTs, by their very nature, are incomplete and, to a certain degree, inaccurate. Their sole importance lies in the possibility that a match may be found with another EST, or more complete piece of information in the available databases. DNA databases now contain such a rich variety of sequence data (ranging from full-length CDSs to genomic sequences and ESTs) that adding EST data may not seem to constitute a problem. However, when EST data were first incorporated in the public datasets, some consternation was expressed that the quality and validity of these resources was being compromised. This consternation probably arose as a result of unfamiliarity with the techniques required to analyse this new form of data, mixed with a degree of concern that ESTs were a shortcut – and perhaps a rather underhand one at that – to developing a complete gene set, a goal that ought properly to belong to the Human Genome Project.

Whatever the political sensitivities, ESTs do add a factor of noise to databases, if only because there is always some degree of uncertainty as to the accuracy of the single pass sequence. On the positive side, ESTs enrich the DNA sequence databases by adding partial sequence representations of some genes that are not otherwise available in those resources, whether coding or non-coding.

5.10.1 EST analysis tools

There are many tools available for the analysis of ESTs: some of these are only available commercially (e.g., the Incyte LifeTools) and will not be dealt with here. The publicly available tools fall into three categories:

- sequence similarity search tools
- sequence assembly tools
- sequence clustering tools

Sequence similarity search tools
The theory underlying sequence similarity search tools is dealt with in Chapter 6. Here we consider the tools as they relate to ESTs.

Current programs are generally designed to cope with ESTs, whether as the search query itself or as a component of the search database. The BLAST series of programs has variants that will translate DNA databases (TBLASTN), translate the input sequence (BLASTX), or both (TBLASTX). (Note that these terms apply to the version 2 BLAST programs; subsequent versions may implement these searches in a different way – bioinformatics programs are rapidly developing!) These options give the widest latitude in dealing with the inherent fuzziness of EST database searches. FastA provides a similar suite of options.

Sequence assembly tools

When a search of the databases reveals several ESTs matching with a probe sequence, normally the ESTs must be aligned with each other to reveal the consensus sequence (see Chapter 7). Usually, further rounds of searching with the consensus identify additional ESTs that should be incorporated into the alignment. This type of iterative sequence alignment is called sequence assembly. There are various tools available to help with this process, e.g., the Staden assembler, the TIGR assembler, Phrap, etc.

Sequence clustering tools

Sequence clustering tools are programs that take a large set of sequences and divide them into subsets, or clusters, based on the extent of shared sequence identity in a minimum overlap region. A reliable and effective mechanism for clustering ESTs will reduce redundancy in the dataset, and save database search time and analysis effort. Such tools are particularly valuable when, for example, large numbers of ESTs have been generated and we need to esti-mate how many different genes are represented by the set. The use of EST clustering to address this problem is outlined in Figure 5.6.

One approach to clustering uses known genes to guide the partitioning of the ESTs. The ESTs are searched against a comprehensive range of DNA and protein sequence databases, and the hits are then sorted into sets (often called buckets), representing individual genes. This approach normally

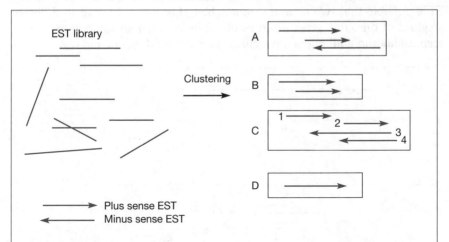

Figure 5.6 Clustering an EST library. A small library of ESTs is illustrated, which, following clustering, is partitioned into four sets, A–D: set A contains three members that exhibit a short overlap, two ESTs overlapping in the plus sense, the third in the minus sense (indicated by arrows); set B contains two members with a substantial overlap in the same sense; set C contains four members, where the two plus-sense ESTs do not overlap with each other, but do overlap with the consensus formed by the two minus-sense members (the first and fourth members also do not overlap with each other, but clearly belong to the same sequence cluster, whose integrity hinges around the third member); and finally, set D is a singleton, having only one member that shares no significant identity with any other EST in the library. The sense of the ESTs is symbolic only – the actual sense can only be determined by database comparison.

yields groups of ESTs that do not match any of the database sequences. Typically, the proportion of ESTs from a given library that remains unmatched following database comparison is ~40%, a value that will decrease as more information becomes available from genome sequencing projects. These remaining sequences must be clustered on the basis of overlapping identities, as illustrated in Figure 5.6.

The alternative strategy is to cluster all the ESTs, generate a consensus sequence representing each cluster, and then perform database searches using only the cluster consensus sequences. This is the ideal solution, as it significantly reduces the number of database similarity searches that have to be performed. However, the success of this strategy depends on how reliably ESTs can be clustered, which in turn depends on the quality of the EST data.

A further complication arises if we wish to estimate the number of genes represented in a library of ESTs, because the unmatched ESTs may not all represent different genes. Two cases should be considered. In the first case, illustrated in Figure 5.7(a), a cluster (C in the figure) might map to an uncharacterised portion of a gene, the characterised portion of which has already been matched by a set, or sets, of ESTs (A and B) (e.g., the 3' UTR is often incomplete in database entries, or a gene sequence may be only a partial sequence entry). In this case, counting the unmatched EST cluster as a representative of a separate gene will bias the gene count towards too high a number. In the second case, illustrated in Figure 5.7(b), it is possible that two or more unmatched clusters could map to different regions of the same gene, again resulting in too high an estimation of gene representation if all unmatched clusters are counted independently.

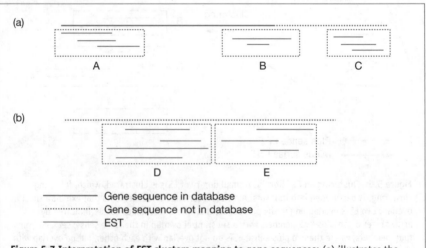

Figure 5.7 Interpretation of EST clusters mapping to gene sequences: (a) illustrates the situation in which EST clusters A and B match an incomplete database entry, and cluster C falls in the uncharacterised region; (b) illustrates the situation in which clusters D and E both map to the same gene, but in a region that has not yet been sequenced and entered into the database.

5.11 A practical example of EST analysis

In this example, we ask the following question of the dbEST EST sequence database: 'Given the availability of the human sequences for cyclooxygenase-1 (COX-1) and cyclooxygenase-2 (COX-2), are there any other close family members?' The drug discovery context for such a question is that non-steroidal anti-inflammatory drugs (NSAIDs), such as aspirin, cause gastric lesions and other, less common, side effects when used in long-term therapy. This results from interference of NSAIDs with cyclooxygenase, an enzyme involved in regulation of the gastric mucosa. The subsequent discovery that there are two cyclooxygenases, designated COX-1 (which protects the gastric mucosa) and COX-2 (which is involved in inflammatory

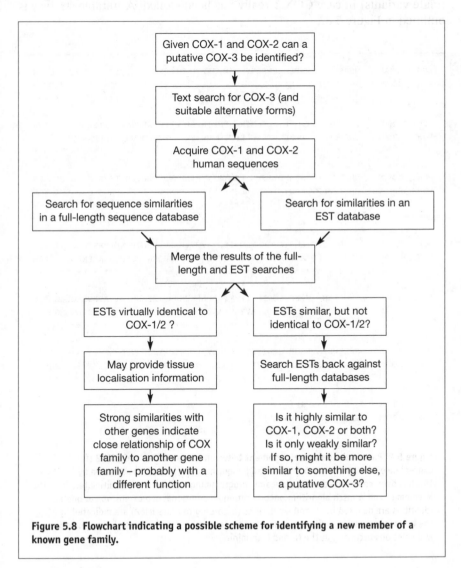

Figure 5.8 Flowchart indicating a possible scheme for identifying a new member of a known gene family.

responses throughout the body), has led to the development of compounds that selectively inhibit one enzyme or the other (for a review see Needleman and Isakson, 1998). A potential therapeutic opportunity would exist if a third member of the COX family were to be cloned, because it may be easier to target a putative COX-3 and avoid interacting with COX-1 altogether.

As sequence analysts, we can begin exploring by bringing together the sequences for human COX-1 and COX-2, and performing database searches on (a) a database of full-length sequences (e.g., GenBank), and (b) a database of ESTs (e.g., dbEST). It is in the newly sequenced ESTs that discoveries are likely to be made (full-length sequences are often well characterised by the authors prior to database submission). It is also wise, in such a strategy, to perform a text query with the term COX-3 (and appropriate variants) in case COX-3 really has been cloned! A suitable strategy is outlined in Figure 5.8.

```
PGH1_HUMAN    MSRSLLLRFLLFLLLLPPLPVLLADPGAPTPVNPCCYYPCQHQGICVRFGLDRYQCDCTR
PGH2_HUMAN    ----MLARALLLCAVL----------ALSHTANPCCSHPCQNRGVCMSVGFDQYKCDCTR
              :*  *  **:  :*           .  . ..****  :***::*:*:  .*:*:*:*****


PGH1_HUMAN    TGYSGPNCTIPGLWTWLRNSLRPSPSFTHFLLTHGRWFWEFVN-ATFIREMLMRLVLTVR
PGH2_HUMAN    TGFYGENCSTPEFLTRIKLFLKPTPNTVHYILTHFKGFWNVVNNIPFLRNAIMSYVLTSR
              **:  * **:  *  :  *  ::   *:*:*,  .*::*** :   **:.**   .*:*: :*   *** *


PGH1_HUMAN    SNLIPSPPTYNSAHDYISWESFSNVSYYTRILPSVPKDCPTPMGTKGKKQLPDAQLLARR
PGH2_HUMAN    SHLIDSPPTYNADYGYKSWEAFSNLSYYTRALPPVPDDCPTPLGVKGKKQLPDSNEIVGK
              *:** ******:  :.* ***:***:***** **.** .*****:*.********:: :. :

.....


PGH1_HUMAN    ALVDAFSRQIAGRIGGGRNMDHHILHVAVDVIRESREMRLQPFNEYRKRFGMKPYTSFQE
PGH2_HUMAN    QFVESFTRQIAGRVAGGRNVPPAVQKVSQASIDQSRQMKYQSFNEYRKRFMLKPYESFEE
               :*::*:*******:.****:    : :*:   *  :**:*: *.******** :*** **:*


PGH1_HUMAN    LVGEKEMAAELEELYGDIDALEFY**PG**LLLEKCHPNSIFGESMIEIGAPFSLKGLLGNPIC
PGH2_HUMAN    LTGEKEMSAELEALYGDIDAVELY**PA**LLVEKPRPDAIFGETMVEVGAPFSLKGLMGNVIC
              *.*****:**** *******:*.:**  **.:** :*:.****:*:*:*.********:.** **


PGH1_HUMAN    SPEYWKPSTFGGEVGFNIVKTATLKKLVCLNTKTCPYVSFRVPD-----------ASQ-
PGH2_HUMAN    SPAYWKPSTFGGEVGFQIINTASIQSLICNNVKGCPFTSFSVPDPELIKTVTINASSSRS
              ** *************:::**:::.*:*  *.*  **:.** ***       :*:


PGH1_HUMAN    --DDG-PAVERP--STEL
PGH2_HUMAN    GLDDINPTVLLKERSTEL
              **  *.*    ****
```

Figure 5.9 Part of the pairwise alignment between human COX-1 and COX-2 (for convenience, the central portion of the alignment has been excised, as denoted by '....'). This has been constructed using a dynamic programming method, derived ultimately from the Needleman and Wunsch algorithm, incorporating an additional gap extension parameter. Identities are denoted by '*' and similarities (according to BLOSUM62) are indicated by ':'. The two sequences are similar along most of their length, with the most obvious insertions and deletions occurring at the N- and C-termini.

The COX-1 and COX-2 human sequences are shown as a pairwise alignment in Figure 5.9. It is worth becoming familiar with the overall distribution of identities and similarities between two sequences in a family (or, if sufficient data are available, constructing a multiple sequence alignment (see Chapter 7)) before attempting to interpret database search results.

If we follow the strategy outlined in Figure 5.8, and we discover no text-based matches with 'COX-3' and find no full-length sequence matches in GenBank, then we turn to EST searches. When performing these searches, we can restrict the target dataset to human ESTs in the first instance; only when these have been exhausted do we need to turn to another organism, since the presence of a COX-3 candidate in, say, mouse is a strong indicator for its presence in the human genome. Variation in gene sequence between species, however, makes the task of detecting new family members very difficult.

Consider a strong match to a human sequence, as illustrated in Figure 5.10. The match is identical over its entire length except for two residues near the beginning, where Pro-Gly in the query has been replaced by Leu-Val in the EST. In considering whether this difference is significant, we should take into account the spread of identities and similarities between COX-1 and COX-2; in this region, the sequences are similar, but not identical; yet in the retrieved match, the sequences are virtually identical. The location of the difference is also significant, occurring as it does close to the beginning of the EST (ESTs generally have lower-quality sequence removed from either end, and it could be that this difference is due to poor-quality sequence remaining). In addition, for each of the substitutions, single base changes would result in the differences between the human and the observed EST read (i.e., Pro to Leu, Gly to Val).

Although it is conceivable that a new family member could be discovered with a very high level of identity to only one member of that family, it is more likely that the sequence we are searching for will have the same overall level of similarity as currently known members. One possibility is

```
>gb|R96180|R96180 yt84f11.r1 Homo sapiens cDNA clone 231021 5' similar to gb:M59979
            PROSTAGLANDIN G/H SYNTHASE 1 PRECURSOR (HUMAN);.
            Length = 355

 Score = 203 bits (511), Expect = 4e-51
 Identities = 96/98 (97%), Positives = 96/98 (97%)

Query: 502 FYPGLLLEKCHPNSIFGESMIEIGAPFSLKGLLGNPICSPEYWKPSTFGGEVGFNIVKTA 561
           FY  LLLEKCHPNSIFGESMIEIGAPFSLKGLLGNPICSPEYWKPSTFGGEVGFNIVKTA
Sbjct: 2   FYLVLLLEKCHPNSIFGESMIEIGAPFSLKGLLGNPICSPEYWKPSTFGGEVGFNIVKTA 181

Query: 562 TLKKLVCLNTKTCPYVSFRVPDASQDDGPAVERPSTEL 599
           TLKKLVCLNTKTCPYVSFRVPDASQDDGPAVERPSTEL
Sbjct: 182 TLKKLVCLNTKTCPYVSFRVPDASQDDGPAVERPSTEL 295
```

Figure 5.10 Human EST match with human COX-1. Note the difference of two residues near the beginning of the match.

```
(a)
>gb|T29235|T29235 EST73861 Homo sapiens cDNA 5' end similar to cyclooxygenase 2
          (HT:264).
          Length = 257

 Score = 109 bits (270), Expect = 7e-23
 Identities = 45/71 (63%), Positives = 56/71 (78%)

Query: 513 PNSIFGESMIEIGAPFSLKGLLGNPICSPEYWKPSTFGGEVGFNIVKTATLKKLVCLNTK 572
           P++IFGE+M+E G+PFS KGL+GN IC P YWKPSTFGGEVGF I+ T + + L+C N K
Sbjct: 3   PDAIFGETMVEXGSPFSXKGLMGNVICXPAYWKPSTFGGEVGFQIINTXSXQSLICNNVK 182

Query: 573 TCPYVSFRVPD 583
           CP+ SF VPD
Sbjct: 183 GCPFTSFSVPD 215

(b)
>gb|T29235|T29235 EST73861 Homo sapiens cDNA 5' end similar to cyclooxygenase 2
          (HT:264).
          Length = 257

 Score = 166 bits (415), Expect = 7e-40
 Identities = 78/85 (91%), Positives = 80/85 (93%)

Query: 500 PDAIFGETMVEVGAPFSLKGLMGNVICSPAYWKPSTFGGEVGFQIINTASIQSLICNNVK 559
           PDAIFGETMVE G+PFS KGLMGNVIC PAYWKPSTFGGEVGFQIINT S QSLICNNVK
Sbjct: 3   PDAIFGETMVEXGSPFSXKGLMGNVICXPAYWKPSTFGGEVGFQIINTXSXQSLICNNVK 182

Query: 560 GCPFTSFSVPDPELIKTVTINASSS 584
           GCPFTSFSVPDPELIKTVTI+ASSS
Sbjct: 183 GCPFTSFSVPDPELIKTVTISASSS 257
```

Figure 5.11 EST hit interpretation. Comparison of a human EST hit with (a) human COX-1 and (b) human COX-2.

that the EST represents a polymorphism. It seems, then, that we can probably disregard this match. Another human match in the hitlist seems more promising at first sight, but turns out to be an EST with known similarity to COX-2, as illustrated in Figure 5.11.

Clearly, searching for new members of known gene families is a painstaking process and this section has only scratched the surface of what can be done. Experience in examining alignments returned from database searches is ultimately the best guide.

5.12 Summary

- Sequence comparisons are more sensitive at the protein level, because the redundant genetic code is reduced to a unique set of amino acids; but this loss of degeneracy means that information that relates directly to evolutionary processes is lost.

- DNA sequence databases include genomic sequence data, and hence contain an assortment of data types that cannot be treated alike (e.g., untranslated sequences (UTRs), introns and exons, mRNA, cDNA and translations). This affects the manner in which searches must be interpreted.

- UTRs flank coding regions of RNA or DNA, but are not themselves translated. Translation of DNA to protein via a genetic code table is termed conceptual, indicating that it carries no experimental validation.

- In an arbitrary length of DNA, it is not known which base marks the start of the coding sequence (CDS), so a six-frame translation must be carried out. The challenge is then to determine which is the correct reading frame.

- Features used to indicate possible coding regions in DNA include: sufficient ORF length; presence of flanking Kozak sequences; patterns of codon usage; third base preference; and presence of ribosome binding sites (Shine–Dalgarno sequences) upstream of the start codon.

- The presence of introns and exons in eukaryotic genes can give rise to gene products of different lengths, because not all exons may be included in the final transcript. Resulting proteins are known as splice variants or alternatively spliced forms.

- Complete CDSs are rarely sequenced in one reaction, so variable-length, overlapping fragments are aligned in order to build a consensus – this is sequence assembly. Multiple base reads per position in the sequence give higher confidence in the result.

- A substantial proportion of currently available DNA data comes from Expressed Sequence Tags (ESTs), which are partial sequences. EST production is highly automated and results are often contaminated with ambiguous or missing bases. This gives rise to difficulties in sequence interpretation.

- The hierarchy of genomic information (the chromosomal genome, the expressed genome, the proteome, etc.) requires different analytical tools and interpretative skills to be brought to bear at each level.

- Various approaches to establishing EST libraries have been developed for academic or commercial exploitation. Suppliers of EST information include Merck/IMAGE, Incyte and TIGR.

- Publicly available tools for EST analysis include those for sequence searching, assembly and clustering.

5.13 Further reading

BROWN, T.A. (1994) *DNA Sequencing*. IRL Press, Oxford.

KOZAK, M. (1991) Structural features in eukaryotic mRNAs that modulate the initiation of translation. *Journal of Biological Chemistry*, **266**(30), 19867–19870.

NEEDLEMAN, P. and ISAKSON, P.C. (1998) Selective inhibition of cyclooxygenase 2. *Science and Medicine*, **5**(1), 26–35.

SHARP, P.A. (1994) Split genes and RNA splicing. *Cell*, **77**, 805–815.

CHAPTER SIX

Pairwise alignment techniques

6.1 Introduction

This chapter introduces the concepts of sequence identity, similarity and homology as they apply to the comparison of two sequences, be they protein, DNA or RNA. Pairwise comparison is a fundamental process in sequence analysis, underpinning, as it does, database search algorithms, which seek out relationships based on sequence properties, rather than on simple interrogation of textual annotation. We consider the definition of local and global similarity, and examine the algorithms that fall into these two categories.

6.2 Database searching

Database interrogation can take the form of text queries (for example, 'display all the human adrenergic receptors') or sequence similarity searches (for example, 'given the sequence of a human adrenergic receptor, display all similar sequences in the database'). This is a useful exercise to try, if the reader has little experience in database interrogation, as these two questions produce quite different results!

Text-based querying has its place in the armoury of sequence analysis software and should certainly not be overlooked in any worthwhile analysis. However, research problems typically take the form of a fragment of sequence that bears no textual annotation with which to frame a query. It is the purpose of this section to describe how the relationships between a query sequence (commonly termed the **probe**) and another sequence (often termed the **subject**) can be quantified and their similarity assessed.

In order to identify an evolutionary (homologous) relationship between a newly determined sequence and a known gene family, we need to assess the extent of shared similarity. Where the degree of similarity is low, the relationship must remain putative, until additional evidence has been gathered (for example, through the assessment of biological data or the use of a phylogeny package, or both). Family relationships are important because they allow us to find some order in the apparent chaos that constitutes the genome. From that order derives our ability to make predictions regarding the completeness of families; specifically, if, in a gene family, 10 members are known in rat and only seven have been identified in human, it is highly likely that at least three human members still remain to be discovered. From a pharmacological point of view, this is very valuable information, because it can be used to bolster the evidence for some known biological response, but for which no receptor has yet been identified. From a molecular biological point of view, it represents an opportunity for the cloning of novel human receptors (albeit in a known gene family) using the rat data as template.

In order to be able to assess effectively the results of database searches, we need to have a firm understanding of the way in which the tools work.

6.3 Alphabets and complexity

A sequence consists of letters selected from an alphabet. The complexity of the alphabet is defined to be the number of different letters it contains. For example, the complexity of the alphabet of the English language is 26; for DNA, the complexity of the alphabet is 4; for EST work, 5; and for proteins, 20. Sometimes additional characters are used in an alphabet to indicate a degree of ambiguity in the identity of a particular residue or base. For example, X is a frequent additional character in protein sequences, indicating an unknown residue; B denotes either asparagine or aspartate (Asx), and Z denotes either glutamine or glutamate (Glx). Some alignment programs will deal with these characters as they stand; others will simply replace such ambiguities with a dummy character, reintroducing the ambiguity before printing the result; and still others fail completely to run with this type of input. For the remainder of the chapter, we will focus on protein sequences with a standard 20-character alphabet.

6.4 Algorithms and programs

It is important to note the difference between an algorithm and a program: the former is a set of steps that define some computational process at an abstract level; the latter is the *implementation* of an algorithm. There may be many different implementations of the same algorithm, but these should give the same results, if the algorithm has been clearly defined. However, like recipes, algorithms are open to interpretation, and those published in scientific papers are no exception.

6.5 Comparing two sequences – a simple case

We are now in a position to consider how to develop an algorithm (a recipe) for determining the similarity between two sequences, each selected from an alphabet of complexity 20. The naïve approach is to line up the sequences against each other and insert additional characters to bring the two strings into vertical alignment, as shown in Figure 6.1.

```
                   Unaligned
                   Sequence 1 (query)    AGGVLIIQVG
                                         ||||||
                   Sequence 2 (subject)  AGGVLIQVG

                   Aligned
                   Sequence 1 (query)    AGGVLIIQVG
                                         |||||| ||||
                   Sequence 2 (subject)  AGGVL-IQVG
```

Figure 6.1 Illustration of the use of a gap character '-' to bring two sequences into alignment; vertical bars denote identical matches – six in the first alignment, nine in the second.

At this point, the task is complete. We could score the alignment by counting how many positions match identically at each position; here, the unaligned score is 6, while the aligned score is 9.

In this elementary example, we can see that the score increases when more identical residues have been aligned. But this is only a simple illustration: the sequences are very short (most protein sequences would be 200 to 500 residues long, or more); the sequences are almost the same length (this is rarely the case in real examples); and the sequences are nearly identical anyway (no other arrangement of residues is possible that achieves the optimal score).

The process of alignment can be measured in terms of the number of gaps introduced and the number of mismatches remaining in the alignment. A metric relating such parameters represents the distance between two sequences (the so-called edit distance). Several metrics exist, and different implementations of similar algorithms may use different distance measures to compute and score alignments.

6.6 Sub-sequences

Consider a more realistic pair of sequences. Sequence A is 400 residues long and B contains 650 residues. If sequence A is in its entirety identical to any portion of sequence B, then A is said to be a sub-sequence of B. Gaps simply need to be inserted, as required, to bring A into register with B, as shown in Figure 6.2(a).

Now consider that sequence A has two extended regions that show identity to sequence B. We would need to identify these regions and then insert gaps into A to bring them into alignment with B, as seen in Figure 6.2(b). Our algorithm could stop at that point, having found the highest scoring sub-sequences between A and B. This is an example of a simple

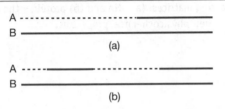

Figure 6.2 Illustration of the alignment of a sub-sequence A with a full-length sequence B, showing: (a) the situation where A is identical to one part of B, and insertion of one block of gaps allows complete alignment of the two sequences; and (b) the situation where A is identical to different parts of B, so that more than one block of gaps must be inserted to bring the sequences into register.

heuristic algorithm that is straightforward to implement, where the regions of identity are obvious.

6.7 Identity and similarity

If sequence comparison depended only on finding regions of strict identity between two sequences, we could develop this method into a reasonable program. However, generally, alignment is not restricted to sub-sequence matching, but involves comparison of full-length sequences. A comprehensive alignment must account fully for the positions of all residues in both sequences. This means that many residues may have to be placed at positions that are not strictly identical. In this case, the positioning of gaps in the alignment becomes more complex to compute. We could simply maximise the number of identical matches by inserting gaps in an unrestricted manner. However, although achieving the optimum score, the result of such a process would be biologically meaningless. Instead, scoring **penalties** are introduced to minimise the number of gaps that are begun (opened), and extension penalties are then incurred when a gap has to be extended. The total alignment score is then a function of the identity between aligned residues and the gap penalties incurred.

In calculating the score for an alignment, we have, thus far, only considered residue identities. Essentially, we have been using a unitary matrix, i.e., one that weights identical residue matches with a score of 1 (see Table 6.1). Such a matrix is termed **sparse**, as most of its elements are zero. This means that its diagnostic power is relatively poor, because all identical matches carry equal weighting. In order to improve diagnostic performance, we want to be able to enhance the scoring potential of the weak, but biologically significant, signals so that they can contribute to the matching process, but without also amplifying noise. This strikes at the very heart of sequence analysis – the imperative to distinguish between high-scoring matches that have only mathematical significance and lower-scoring matches that are biologically meaningful.

Table 6.1 Unitary scoring matrices: (a) DNA and (b) protein – the amino acids are grouped according to their physicochemical properties.

(a)

	A	C	G	T
A	1	0	0	0
C	0	1	0	0
G	0	0	1	0
T	0	0	0	1

(b)

	C	S	T	P	A	G	N	D	E	Q	H	R	K	M	I	L	V	F	Y	W	B	Z	X
C	1	0	0	0	0	0	0	0	0	0	0	0	0	0	0	0	0	0	0	0	0	0	0
S	0	1	0	0	0	0	0	0	0	0	0	0	0	0	0	0	0	0	0	0	0	0	0
T	0	0	1	0	0	0	0	0	0	0	0	0	0	0	0	0	0	0	0	0	0	0	0
P	0	0	0	1	0	0	0	0	0	0	0	0	0	0	0	0	0	0	0	0	0	0	0
A	0	0	0	0	1	0	0	0	0	0	0	0	0	0	0	0	0	0	0	0	0	0	0
G	0	0	0	0	0	1	0	0	0	0	0	0	0	0	0	0	0	0	0	0	0	0	0
N	0	0	0	0	0	0	1	0	0	0	0	0	0	0	0	0	0	0	0	0	0	0	0
D	0	0	0	0	0	0	0	1	0	0	0	0	0	0	0	0	0	0	0	0	0	0	0
E	0	0	0	0	0	0	0	0	1	0	0	0	0	0	0	0	0	0	0	0	0	0	0
Q	0	0	0	0	0	0	0	0	0	1	0	0	0	0	0	0	0	0	0	0	0	0	0
H	0	0	0	0	0	0	0	0	0	0	1	0	0	0	0	0	0	0	0	0	0	0	0
R	0	0	0	0	0	0	0	0	0	0	0	1	0	0	0	0	0	0	0	0	0	0	0
K	0	0	0	0	0	0	0	0	0	0	0	0	1	0	0	0	0	0	0	0	0	0	0
M	0	0	0	0	0	0	0	0	0	0	0	0	0	1	0	0	0	0	0	0	0	0	0
I	0	0	0	0	0	0	0	0	0	0	0	0	0	0	1	0	0	0	0	0	0	0	0
L	0	0	0	0	0	0	0	0	0	0	0	0	0	0	0	1	0	0	0	0	0	0	0
V	0	0	0	0	0	0	0	0	0	0	0	0	0	0	0	0	1	0	0	0	0	0	0
F	0	0	0	0	0	0	0	0	0	0	0	0	0	0	0	0	0	1	0	0	0	0	0
Y	0	0	0	0	0	0	0	0	0	0	0	0	0	0	0	0	0	0	1	0	0	0	0
W	0	0	0	0	0	0	0	0	0	0	0	0	0	0	0	0	0	0	0	1	0	0	0
B	0	0	0	0	0	0	0	0	0	0	0	0	0	0	0	0	0	0	0	0	1	0	0
Z	0	0	0	0	0	0	0	0	0	0	0	0	0	0	0	0	0	0	0	0	0	1	0
X	0	0	0	0	0	0	0	0	0	0	0	0	0	0	0	0	0	0	0	0	0	0	1

To address this problem, scoring matrices have been devised that weight matches between non-identical residues, according to observed substitution rates across large evolutionary distances. Judicious use of such tools may increase the sensitivity of the alignment process, especially in situations where absolute sequence identity is low. However, it must be appreciated that similarity matrices are inherently noisy, because they indiscriminately weight relationships that may be inappropriate in the context of any particular sequence comparison (this means that scores of random matches are boosted along with those of weak signals). It is beyond the scope of this book to discuss exhaustively the numerous matrices available. Instead we focus on two of the most popular series: the original Dayhoff mutation data (MD) and BLOSUM matrices.

6.7.1 The Dayhoff Mutation Data Matrix

The MD (Dayhoff *et al.*, 1978) score is based on the concept of the Point Accepted Mutation (PAM). An evolutionary distance of 1 PAM indicates the probability of a residue mutating during a distance in which 1 point mutation was accepted per 100 residues. Mutation probability matrices corresponding to larger intervals of evolutionary distance can be obtained by repeatedly multiplying the original matrix by itself. The 250 PAM matrix gives similarity scores equivalent to 20% matches remaining between two sequences.

For general sequence comparison, it is useful to employ a matrix whose elements reflect the ratio of the probability of an amino acid exchange to the probability of the two amino acids occurring at random. These ratios are expressed by the elements of the relatedness odds matrix. When one protein is compared with another, position by position, the odds for each position are multiplied to calculate a score for the entire alignment. However, it is computationally more convenient to add the logarithms of the odds. Thus, the MD matrix contains the logs of the elements of the 250 PAM odds matrix (a log odds matrix). As shown in Table 6.2, within the matrix, values greater than 0 indicate likely mutations, values equal to 0 are neutral (random), and values less than 0 indicate unlikely mutations.

Table 6.2 Mutation Data Matrix for 250 PAMs. The amino acids are arranged by assuming that positive values represent evolutionarily conservative replacements. The amino acids are ranked here according to groups based on their physicochemical properties.

	C	S	T	P	A	G	N	D	E	Q	H	R	K	M	I	L	V	F	Y	W
C	12																			
S	0	2																		
T	-2	1	3																	
P	-3	1	0	6																
A	-2	1	1	2																
G	-3	1	0	-1	1	5														
N	-4	1	0	-1	0	0	2													
D	-5	0	0	-1	0	1	2	4												
E	-5	0	0	-1	0	0	1	3	4											
Q	-5	-1	-1	0	0	-1	1	2	2	4										
H	-3	-1	-1	0	-1	-2	2	1	1	3	6									
R	-4	0	-1	0	-2	-3	0	-1	-1	1	2	6								
K	-5	0	0	-1	-1	-2	1	0	0	1	0	3	5							
M	-5	-2	-1	-2	-1	-3	-2	-3	-2	-1	-2	0	0	6						
I	-2	-1	0	-2	-1	-3	-2	-2	-2	-2	-2	-2	-2	2	5					
L	-6	-3	-2	-3	-2	-4	-3	-4	-3	-2	-2	-3	-3	4	2	6				
V	-2	-1	0	-1	0	-1	-2	-2	-2	-2	-2	-2	-2	2	4	2	4			
F	-4	-3	-3	-5	-4	-5	-4	-6	-5	-5	-2	-4	-5	0	1	2	-1	9		
Y	0	-3	-3	-5	-3	-5	-2	-4	-4	-4	0	-4	-4	-2	-1	-1	-2	7	10	
W	-8	-2	-5	-6	-6	-7	-4	-7	-7	-5	-3	2	-3	-4	-5	-2	-6	0	0	17

Our aim in sequence analysis is to identify relationships in the Twilight Zone (see Chapter 1). The MD for 250 PAMs has thus become the default matrix in many analysis packages, because it reflects identities at the 20% level. However, in principle, it is more effective to use a matrix that corresponds to the actual evolutionary distance between the sequences being compared (see Figure 6.3) – this makes sense, but is limiting because it requires *a priori* knowledge of that distance (in other words, it requires us to know in advance what we are looking for!). It is therefore good practice to establish a strategy in which applications are run, and results assessed, using a range of different PAM matrices.

OBSERVED % DIFFERENCE	EVOLUTIONARY DISTANCE (PAMs)
1	1
10	11
20	23
30	38
40	56
50	80
60	112
70	159
80	246

Figure 6.3 Correspondence between observed residue identity differences and evolutionary distance (measured in PAMs).

6.7.2 The BLOSUM matrices

Matrices based on the Dayhoff model of evolutionary rates are of limited value because their substitution rates are derived from alignments of sequences that are at least 85% identical. However, as we have seen, the most common task in sequence analysis is the detection of more distant relationships, which are only *inferred* from the Dayhoff model.

Recognising these limitations, Henikoff and Henikoff (1992) derived a set of substitution matrices from blocks of aligned sequences in the BLOCKS database, in order to represent distant relationships more explicitly. In deriving the matrices, any bias potentially introduced by counting multiple contributions from identical residue pairs is removed by clustering sequence segments on the basis of minimum percentage identity. For each cluster, the average contribution at each residue position is calculated, so that effectively the clusters are treated as single sequences. Different matrices then emerge by setting different clustering percentages. Thus, for example, sequences clustered at greater than or equal to 80% identity are used to generate the BLOSUM 80 matrix (BLOcks SUbstitution Matrix – pronounced blossom); those in the 62% or greater cluster contribute to the BLOSUM 62 matrix, and so on.

Figure 6.4 illustrates pairwise comparisons carried out using the PAM 250 and BLOSUM 62 matrices, where it can be observed that different residue relationships are considered to be significant when the alternative matrices are in operation. For example, at the start of these alignments, PAM 250 considers a K/E substitution unlikely, whereas BLOSUM 62 favours the match; within the body of the alignments, G/A and G/S substitutions considered favourable by PAM 250 are not scored by BLOSUM 62; and at the end of the alignments, an E/K substitution, disregarded by PAM 250, results in an extension of the alignment by one residue when BLOSUM 62 is used. These are seemingly trivial differences, because we are comparing two highly similar sequences, but may be crucial considerations in the Twilight Zone, where the enhancement and ultimate detection of the weakest of signals is all-important.

```
(a)
Identities = 36/52 (69%), Positives = 47/52 (90%)

Query:  214  KMGPGFTKALGHGVDLGHIYGDNLERQYQLRLFKDGKLKYQVLDGEMYPPSV  265
                 GP+FTK+   HGVDL+HIYG++LERQ +LRLFKDGK+KYQ+++GEMYPP+V
Sbjct:   97  ERGPAFTKGKNHGVDLSHIYGESLERQHKLRLFKDGKMKYQMINGEMYPPTV  148
(b)

Identities = 36/53 (68%), Positives = 47/53 (89%)

Query:  214  KMGPGFTKALGHGVDLGHIYGDNLERQYQLRLFKDGKLKYQVLDGEMYPPSVE  266
               + GP FTK    HGVDL HIYG++LERQ++LRLFKDGK+KYQ+++GEMYPP+V+
Sbjct:   97  ERGPAFTKGKNHGVDLSHIYGESLERQHKLRLFKDGKMKYQMINGEMYPPTVK  149
```

Figure 6.4 Pairwise comparisons using different scoring matrices: (a) PAM 250, and (b) BLOSUM 62. The overall results are similar, but differ in the details of which particular residue relationships are considered to be significant. Likely substitutions are indicated by + signs. The 'Identities' keyword indicates the number of identical residue matches in relation to the match length; the 'Positives' keyword considers both identities and similarities.

6.7.3 Statistical measures of alignment significance

When performing sequence alignment computationally, we are really creating a match between two sequences according to a mathematical model. The model describes, in general terms, the concept of alignment of two sequence strings, and the fine detail (gap penalties, impact of sequence length differences, effect of alphabet complexity, and so on) is dealt with through the use of parameters. Appropriate choice of parameters will minimise the number of gaps, while relaxing the parameters will, theoretically, allow alignment of any arbitrary sequences. The fact that a program produces an alignment between two sequences should not be taken as proof in itself that any relationship exists between them.

Any standard program will produce some statistical value indicating the level of confidence that should be attached to an alignment. The statistics quoted in BLAST for pairwise comparisons are probability (p) or expected frequency (E) values. The p-value relates the score returned for an alignment to the likelihood of its having arisen by chance: in general, the closer the value approaches to zero, the greater the confidence that the match is real; conversely, the nearer the value is to unity, the greater the chance that the match is spurious. The E-value (or Expect value) describes the number of hits one can 'expect' to see by chance (in other words noise) when searching a database of a particular size. For example, an E-value of 1 assigned to a hit can be interpreted as meaning that, in the current search, one might expect to see one match with a similar score simply by chance; conversely, a value of zero indicates that no matches would be expected by chance.

Inspection of the alignment in Figure 6.5 indicates a high level of sequence similarity (90%) and a correspondingly low E-value (7e–34), showing that this alignment is unlikely to be a random match.

```
>bbs|69040 70 kda cyclooxygenase-related protein [mice, Peptide Partial, 80 aa]

 Score = 145 bits (362), Expect = 7e-34
 Identities = 66/80 (82%), Positives = 73/80 (90%)


Query: 294  LPGLMLYATLWLREHNRVCDLLKAEHPTWGDEQLFQTTRLILIGETIKIVIEEYVQQLSG 353
            +PGLM+YAT+WLREHNRVCDLLK EHP WGDEQLFQT+RLILIGETIKIVIE+YVQ LSG
Sbjct: 1    VPGLMMYATIWLREHNRVCDLLKQEHPEWGDEQLFQTSRLILIGETIKIVIEDYVQHLSG 60

Query: 354  YFLQLKFDPELLFGVQFQYR 373
            Y +LKFDPELLF QFQY+
Sbjct: 61   YHFKLKFDPELLFNQQFQYQ 80
```

Figure 6.5 Part of a hitlist from a database search indicating the scoring data for one pairwise comparison. Here, the calculated score is shown in bits and the E-value is denoted by 'Expect'.

6.8 The Dotplot

Perhaps the most basic method of comparing two sequences is a visual approach known as a dotplot. Consider two sequences, A and B, whose lengths can be different, but in the ideal case are fairly similar. We proceed by creating a rectangular matrix in which the residues of A are mapped along the x-axis, and those of B along the y-axis. Initially, the matrix is filled with zeros. Each of its cells, $x_i y_j$ (where i varies between 1 and the sequence length of A, and j varies between 1 and the length of B), is considered in turn and is assigned a value indicating the level of similarity between the two residue positions (A_i and B_j). In the simplest scheme, all cells remain zero, unless $A_i = B_j$, in which case the element is assigned the value 1.

Such a matrix can be visualised quite simply for short sequences, for example by printing out the matrix in a fixed-pitch font, as shown in

Table 6.3; or for longer sequences, by using an appropriate graphics program, as shown in Figure 6.6. The plot is characterised by some apparently random dots (noise) and a central diagonal line, where a high density of adjacent dots (the signal) indicates the regions of greatest similarity between the two sequences (for clarity, dots are marked in Table 6.3 as Xs). The dotplot is effectively a signal-to-noise landscape, which offers a tangible, graphic expression of the challenge of discriminating signal from noise. It is often helpful to recall this image when interpreting database search results.

Within a dotplot, two *identical* sequences are characterised by a single unbroken diagonal line across the plot, as shown graphically in Figure 6.6(a). Note that, for full-length sequences, a plot must be reduced in size in order to be able to visualise the complete comparison – in so doing, the Xs shown in the magnified section shown in Table 6.3 are reduced to dots (hence dotplot), which, at sufficiently low magnification, will ultimately merge into lines.

By contrast with identical sequences, two *similar* sequences will be characterised by a broken diagonal, the interrupted region indicating the

Table 6.3 Illustration of the manner of construction of the dotplot matrix, using a simple residue identity matrix to score an 'X' where a pair of identical residues is observed.

	M	T	F	R	D	L	L	S	V	S	F	E	G	P	R	P	D	S	S	A	G	G	S	S	A	G	G
M	X																										
T		X																									
F			X								X																
R				X											X												
D					X												X										
L						X	X																				
L						X	X																				
S								X		X								X	X				X	X			
V									X																		
S								X		X								X	X				X	X			
F			X								X																
E												X															
G													X								X	X				X	X
P														X		X											
R				X											X												
P														X		X											
D					X												X										
S								X		X								X	X				X	X			
S								X		X								X	X				X	X			
A																				X					X		
G													X								X	X				X	X
G													X								X	X				X	X

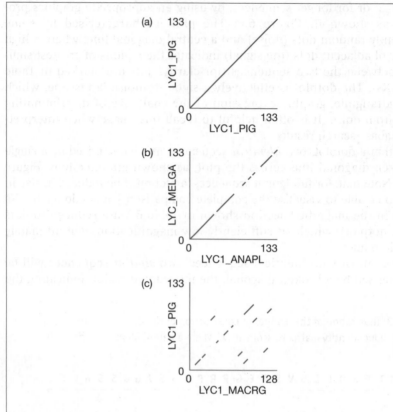

Figure 6.6 Graphical representation of dotplots, showing comparisons of (a) two identical sequences; (b) two highly similar sequences; and (c) two different, but related sequences.

location of sequence mismatches, as shown in Figure 6.6(b). A pair of distantly related sequences, with fewer similarities, has a much noisier plot; in this case, diagonal clusters of dots are observed, parallel to the central diagonal, separated by a distance that represents the number of insertions required to bring the sequences into the correct register – see Figure 6.6(c).

Until now, we have been discussing the construction of dotplots using a unitary matrix. More sophisticated dotplots exploit advanced scoring schemes for calculating cell scores. For example, the score for each pair of residues may be assigned with reference to different types of similarity matrix, which might allow comparison on the basis of evolutionary relatedness, structural similarity, physicochemical properties, and so on. In such cases, it becomes more important to filter out noise, because the plots no longer simply result from the application of a sparse matrix, and off-diagonal components are receiving higher weighting through amplification of both signal and background noise. The practical solution to this is to implement a sliding window calculation, as a smoothing function, to improve the signal-to-noise ratio.

We have already seen that alignments are simply mathematical models whose behaviour can be modified through the use of parameters. Different models exist, which are designed to encapsulate a variety of physical characteristics of biological sequences, including, for example, their structural, functional and/or evolutionary relatedness. In this context, we should therefore be aware that there is no right or wrong alignment, but rather different models reflecting different biological perspectives.

Two general models view alignments in rather different ways: the first considers similarity across the full extent of the sequences (a global alignment); the second focuses on regions of similarity in parts of the sequences only (a local alignment). It is important to understand these distinctions, to appreciate that sequences are not uniformly similar, and that there is therefore no value in performing a global alignment on sequences that have only local similarity.

Some of the publicly available implementations of pairwise comparison programs (e.g., BLAST and FastA) are fast because they look for local alignments, and are made to run even faster by incorporating heuristics.

The rationale for local similarity searching is that functional sites (e.g., catalytic sites of enzymes) are localised to relatively short regions, which are conserved irrespective of deletions or mutations in intervening parts of the sequence. Thus, a search for local similarity may produce more biologically meaningful and sensitive results than a search attempting to optimise alignment over the entire sequence lengths. It should be noted that it does not follow that small similar sequence fragments necessarily have the same 3D fold.

6.10 Global alignment: the Needleman and Wunsch algorithm

Having discussed some general concepts, we now turn to the Needleman and Wunsch algorithm for computing a global alignment between two sequences. Conceptually, this method is similar to a dotplot that has been interpreted computationally. In this approach, a maximum match between two sequences is defined to be the largest number of amino acids from one protein that can be matched with those of another protein, while allowing for all possible deletions. A penalty is introduced, to provide a barrier to arbitrary gap insertion.

As before with the diagonal plot, sequences are compared by constructing a 2D matrix. Needleman and Wunsch (1970) proposed a maximum-match pathway that can be obtained computationally by applying a straightforward algorithm. In its simplest form, cells representing identities are scored 1, and cells representing mismatches are scored 0; the 2D array is thus populated with these values. An operation of successive summation of cells then commences. This process examines each cell in the matrix, the maximum score along any path leading to the cell is added to its present contents, and the summation continues. When this process has been completed, the maximum-match pathway is constructed. An alignment is generated by working through the matrix, starting at the highest-scoring element at the N-termini, and following the pattern of high scores through to the C-termini. Leaps to non-adjacent diagonal cells in the matrix indicate the need for gap insertion, to bring the sequences into register, as long as the gap penalty barrier permits opening a gap at that point. If the barrier does not permit gap insertion, a lower-scoring pathway may have to be taken.

Let us consider what this means in more detail by comparing two short sequences, 'ADLGAVFALCDRYFQ' and 'ADLGRTQNCDRYYQ'. Following the algorithm through, step by step, we first construct a 2D matrix and populate it with scores representing the residue identities between the two sequences, as shown in Table 6.4.

Table 6.4 Identity matrix used to initiate the Needleman and Wunsch alignment procedure for the sequences shown. Identical residue pairs are scored 1 in the appropriate matrix cell.

	A	D	L	G	A	V	F	A	L	C	D	R	Y	F	Q
A	1	0	0	0	1	0	0	1	0	0	0	0	0	0	0
D	0	1	0	0	0	0	0	0	0	0	1	0	0	0	0
L	0	0	1	0	0	0	0	0	1	0	0	0	0	0	0
G	0	0	0	1	0	0	0	0	0	0	0	0	0	0	0
R	0	0	0	0	0	0	0	0	0	0	0	1	0	0	0
T	0	0	0	0	0	0	0	0	0	0	0	0	0	0	0
Q	0	0	0	0	0	0	0	0	0	0	0	0	0	0	1
N	0	0	0	0	0	0	0	0	0	0	0	0	0	0	0
C	0	0	0	0	0	0	0	0	0	1	0	0	0	0	0
D	0	1	0	0	0	0	0	0	0	0	1	0	0	0	0
R	0	0	0	0	0	0	0	0	0	0	0	1	0	0	0
Y	0	0	0	0	0	0	0	0	0	0	0	0	1	0	0
Y	0	0	0	0	0	0	0	0	0	0	0	0	1	0	0
Q	0	0	0	0	0	0	0	0	0	0	0	0	0	0	1

Successive summation of cells begins at the last cell in the matrix, which we term the leading cell. In this implementation, the leading cell progresses up the last column, which simply contains the values from the equivalent column of the identity matrix (hence there are two 1's where there are coincident glutamine residues). The program continues from the bottom of the previous column, which starts with a 0 (because F and Q are non-identical); the next cell represents the Y-F pair, which are also non-identical, yet the algorithm scores a 1 because the only path leading to it has a maximum score of 1. The same condition holds for all positions in this column. Moving to the third column, 0 appears in the Q-Y position (because these residues are non-identical); in the Y-Y position, the existing value is 1 because the residues are identical, to which is added the value 1, from the Q-Q position, because this is the maximum value from the only path (the so-called sub-path) leading to it. Hence the score of this Y-Y cell is 2, as shown in Table 6.5.

Table 6.5 Partially complete maximum-match matrix. The operation is about to be carried out on the highlighted cell L-L, by adding the highest value (5 at position C-C) from the two sub-paths leading to it, to give a value of 6.

	A	D	L	G	A	V	F	A	L	C	D	R	Y	F	Q
A	1	0	0	0	1	0	0	1	0	4	3	2	1	1	0
D	0	1	0	0	0	0	0	0	0	4	4	2	1	1	0
L	0	0	1	0	0	0	0	0	1	4	3	2	1	1	0
G	0	0	0	1	0	0	0	0	5	4	3	2	1	1	0
R	0	0	0	0	0	0	0	0	5	4	3	3	1	1	0
T	0	0	0	0	0	0	0	0	5	4	3	2	1	1	0
Q	0	0	0	0	0	0	0	0	5	4	3	2	1	1	1
N	0	0	0	0	0	0	0	0	5	4	3	2	1	1	0
C	0	0	0	0	0	0	0	0	4	5	3	2	1	1	0
D	0	1	0	0	0	0	0	0	3	3	4	2	1	1	0
R	0	0	0	0	0	0	0	0	2	2	2	3	1	1	0
Y	0	0	0	0	0	0	0	0	2	2	2	2	2	1	0
Y	0	0	0	0	0	0	0	0	1	1	1	1	2	1	0
Q	0	0	0	0	0	0	0	0	0	0	0	0	0	0	1

In this way, the algorithm proceeds from column to column, moving deeper into the matrix, and gathering two sub-paths from which to select the highest value to add to the leading cell. By the time we get to the seventh column, in considering the highlighted L-L position, the existing value is 1, and the highest value from either of the two sub-paths is 5; thus, the value of the leading cell here becomes 6 (see Table 6.6). The remainder of the matrix is populated in a similar manner, to give the final matrix illustrated in Table 6.6.

From the complete matrix, a path is traced back from the highest-scoring position (which, by definition, must occur at the N-termini, because of the cumulative nature of the summation process) to the origin. In this example, there is a break in continuity of the path, which can only be reconciled by the insertion of a gap between Asn and Cys in the shorter (vertical) sequence. Piecing together the alignment from this process yields the result depicted in Figure 6.7.

Note that using this scoring scheme, the maximum match value (9), observed at the matrix origin, represents the number of identities lying along the match pathway.

From this description, it is clear that the Needleman and Wunsch algorithm produces an alignment that takes into account all residues of the input sequences. The starting point of the maximum-path traceback is always at the N-termini, and is calculated from a scoring process that commences at the C-termini. For this reason, the method results in a global alignment.

Table 6.6 Completed matrix in which the value being calculated in Table 6.5 is boxed, and the maximum-match pathway giving the highest scoring alignment is highlighted.

	A	D	L	G	A	V	F	A	L	C	D	R	Y	F	Q
A	9	7	6	6	7	6	6	7	5	4	3	2	1	1	0
D	7	8	6	6	6	6	6	6	5	4	4	2	1	1	0
L	6	6	7	5	5	5	5	5	6	4	3	2	1	1	0
G	5	5	5	6	5	5	5	5	5	4	3	2	1	1	0
R	5	5	5	5	5	5	5	5	5	4	3	3	1	1	0
T	5	5	5	5	5	5	5	5	5	4	3	2	1	1	0
Q	5	5	5	5	5	5	5	5	5	4	3	2	1	1	1
N	5	5	5	5	5	5	5	5	5	4	3	2	1	1	0
C	4	4	4	4	4	4	4	4	4	5	3	2	1	1	0
D	3	4	3	3	3	3	3	3	3	3	4	2	1	1	0
R	2	2	2	2	2	2	2	2	2	2	2	3	1	1	0
Y	2	2	2	2	2	2	2	2	2	2	2	2	2	1	0
Y	1	1	1	1	1	1	1	1	1	1	1	1	2	1	0
Q	0	0	0	0	0	0	0	0	0	0	0	0	0	0	1

```
ADLGAVFALCDRYFQ
| | | |       | | | |  |
ADLGRTQN-CDRYYQ
```

Figure 6.7 Final gapped alignment resulting from an implementation of the Needleman and Wunsch algorithm.

6.11 Local alignment: the Smith–Waterman algorithm

The Needleman and Wunsch algorithm just described works well for sequences that show similarity across most of their lengths. Consider, however, two sequences that are only distantly related to each other; they will, even so, exhibit small regions of local similarity, although no satisfactory overall alignment can be found. In 1981, Smith and Waterman described a method, commonly known as the Smith–Waterman algorithm, for finding these common regions of similarity. Like the technique of Needleman and Wunsch, this is a matrix-based approach, and backtracking is used to reconstruct the gapped alignments. The Smith–Waterman method has been used as the basis for many subsequent algorithms, and is often quoted as a benchmark when comparing different alignment techniques. It is certainly a sensitive technique, but it should be remembered that when using any implementation of it, the function of the algorithm is to find small, locally similar regions.

A key feature of the Smith–Waterman algorithm is that each cell in the matrix defines the end point of a potential alignment, whose similarity is represented by the value stored in the cell. The algorithm thus begins by filling the edge elements with 0.0 values, as illustrated in Table 6.7, because these cells represent the ends of alignments of length zero and, consequently, their similarity score is zero. Note that, here, cells in the matrix are populated with floating-point values, rather than integers, which are characteristic of the Needleman and Wunsch method described above; however, there is no reason why either method could not be implemented using integers or floating-point values.

Table 6.7 The starting point for the Smith–Waterman algorithm in which the edge elements are initialised to 0.0. The symbol 'x' is used as a placeholder, because the first row and first column cannot be the endpoint of any alignment.

	x	A	D	L	G	A	V	F	A	L	C	D	R	Y	F	Q
x	0.0	0.0	0.0	0.0	0.0	0.0	0.0	0.0	0.0	0.0	0.0	0.0	0.0	0.0	0.0	0.0
A	0.0															
D	0.0															
L	0.0															
G	0.0															
R	0.0															
T	0.0															
Q	0.0															
N	0.0															
C	0.0															
D	0.0															
R	0.0															
Y	0.0															
Y	0.0															
Q	0.0															

The next step is to populate the remaining cells in the matrix. This is achieved by evaluating three functions and choosing the maximum of the three values, or zero if a negative value would result. These functions consider the possibilities for ending an alignment at any particular cell. First, the similarity score (e.g., 1.0 for a match, –0.333 for a mismatch) for the diagonal predecessor of the cell under consideration is added to that cell's score (see Table 6.8); then the maximum value is calculated for a deletion represented along (a) the current row of the matrix, and (b) along the current column of the matrix. Finally, if a negative score would result, 0.0 is substituted, to indicate that there is no alignment similarity up to the current cell position.

Once the matrix is complete, the highest score is located (representing the endpoint of the highest scoring alignment between the two sequences), and the other elements leading to this cell are determined using a backtracking procedure, as illustrated in Table 6.9. If necessary, we can search the matrix for lower-scoring local alignments simply by finding other high scores that do not form part of a previous traceback.

Table 6.8 Calculation of first set of diagonal similarity scores in the Smith–Waterman algorithm.

	x	A	D	L	G	A	V	F	A	L	C	D	R	Y	F	Q
x	0.0	0.0	0.0	0.0	0.0	0.0	0.0	0.0	0.0	0.0	0.0	0.0	0.0	0.0	0.0	0.0
A	0.0	1.0	0.0	0.0	0.0	1.0	0.0	0.0	1.0	0.0	0.0	0.0	0.0	0.0	0.0	0.0
D	0.0	0.0	2.0	0.0	0.0	0.0	0.7	0.0	0.0	0.7	0.0	1.0	0.0	0.0	0.0	0.0
L	0.0	0.0	0.0	3.0	0.0	0.0	0.0	0.3	0.0	1.0	0.3	0.0	0.7	0.0	0.0	0.0
G	0.0	0.0	0.0	0.0	4.0	0.0	0.0	0.0	0.0	0.0	0.7	0.0	0.0	0.3	0.0	0.0
R	0.0	0.0	0.0	0.0	0.0	3.7	0.0	0.0	0.0	0.0	0.0	0.3	1.0	0.0	0.0	0.0
T	0.0	0.0	0.0	0.0	0.0	0.0	3.3	0.0	0.0	0.0	0.0	0.0	0.0	0.7	0.0	0.0
Q	0.0	0.0	0.0	0.0	0.0	0.0	0.0	3.0	0.0	0.0	0.0	0.0	0.0	0.0	0.3	1.0
N	0.0	0.0	0.0	0.0	0.0	0.0	0.0	0.0	2.7	0.0	0.0	0.0	0.0	0.0	0.0	0.0
C	0.0	0.0	0.0	0.0	0.0	0.0	0.0	0.0	0.0	2.3	1.0	0.0	0.0	0.0	0.0	0.0
D	0.0	0.0	1.0	0.0	0.0	0.0	0.0	0.0	0.0	0.0	2.0	2.0	0.0	0.0	0.0	0.0
R	0.0	0.0	0.0	0.7	0.0	0.0	0.0	0.0	0.0	0.0	0.0	1.7	3.0	0.0	0.0	0.0
Y	0.0	0.0	0.0	0.0	0.3	0.0	0.0	0.0	0.0	0.0	0.0	0.0	1.3	4.0	0.0	0.0
Y	0.0	0.0	0.0	0.0	0.0	0.0	0.0	0.0	0.0	0.0	0.0	0.0	0.0	2.3	3.7	0.0
Q	0.0	0.0	0.0	0.0	0.0	0.0	0.0	0.0	0.0	0.0	0.0	0.0	0.0	0.0	2.0	4.7

Table 6.9 The endpoint of the Smith–Waterman algorithm after calculation of all scoring parameters. A traceback from the highest score is highlighted.

	x	A	D	L	G	A	V	F	A	L	C	D	R	Y	F	Q
x	0.0	0.0	0.0	0.0	0.0	0.0	0.0	0.0	0.0	0.0	0.0	0.0	0.0	0.0	0.0	0.0
A	0.0	1.0	0.0	0.0	0.0	1.0	0.0	0.0	1.0	0.0	0.0	0.0	0.0	0.0	0.0	0.0
D	0.0	0.0	2.0	0.7	0.3	0.0	0.7	0.0	0.0	0.7	0.0	1.0	0.0	0.0	0.0	0.0
L	0.0	0.0	0.7	3.0	1.7	1.3	1.0	0.7	0.3	1.0	0.3	0.0	0.7	0.0	0.0	0.0
G	0.0	0.0	0.3	1.7	4.0	2.7	2.3	2.0	1.7	1.3	1.0	0.7	0.3	0.3	0.0	0.0
R	0.0	0.0	0.0	1.3	2.7	3.7	2.3	2.0	1.7	1.3	1.0	0.7	1.0	0.0	0.0	0.0
T	0.0	0.0	0.0	1.0	2.3	2.3	3.3	2.0	1.7	1.3	1.0	0.7	0.3	0.7	0.0	0.0
Q	0.0	0.0	0.0	0.7	2.0	2.0	2.0	3.0	1.7	1.3	1.0	0.7	0.3	0.0	0.3	1.0
N	0.0	0.0	0.0	0.3	1.7	1.7	1.7	1.7	2.7	1.3	1.0	0.7	0.3	0.0	0.0	0.0
C	0.0	0.0	0.0	0.0	1.3	1.3	1.3	1.3	1.3	2.3	1.0	0.7	0.3	0.0	0.0	0.0
D	0.0	0.0	1.0	0.0	1.0	1.0	1.0	1.0	1.0	1.0	2.0	2.0	0.7	0.3	0.0	0.0
R	0.0	0.0	0.0	0.7	0.7	0.7	0.7	0.7	0.7	0.7	0.7	1.7	3.0	1.7	1.3	1.0
Y	0.0	0.0	0.0	0.0	0.3	0.3	0.3	0.3	0.3	0.3	0.3	0.3	1.7	4.0	2.7	2.3
Y	0.0	0.0	0.0	0.0	0.0	0.0	0.0	0.0	0.0	0.0	0.0	0.0	1.3	2.7	3.7	2.3
Q	0.0	0.0	0.0	0.0	0.0	0.0	0.0	0.0	0.0	0.0	0.0	0.0	1.0	2.3	2.3	4.7

The essential difference between the two algorithms we have examined so far is that, in the Smith–Waterman case, the matrix contains a maximum value that may not be at the N-termini of the sequences. It represents the endpoint of an alignment such that no other pair of segments with greater similarity exists between the two sequences. Hence, this is a local, rather than a global, alignment method.

The methods outlined in the preceding sections fall into the general category of dynamic programming algorithms. This is a programming technique in which we build a solution to a problem by solving smaller but similar sub-problems. As we have discovered, between two sequences that are more than trivially different, there is not simply one (correct) alignment, but often several possible alignments. Finding a good solution using dynamic programming usually involves the technique of backtracking and testing different paths to high-scoring alignments, guided by the various parameters (gap penalties, etc.) available to the algorithm. The best of all paths (i.e., the one that most effectively links together the sub-problems into an optimal solution) is then selected as the final alignment.

6.13 Pairwise database searching

Performing a comparison of one sequence against a database of many thousands can be viewed as simply an extension of pairwise alignment. Achieving a database search in an efficient manner is not trivial, and, as datasets get larger, more effort is being spent to try to improve efficiency. To perform a Needleman and Wunsch, or Smith–Waterman, alignment is practicable for small numbers of sequences, but for large database searches the methods can become prohibitively time-consuming.

Implementations of the Smith–Waterman algorithm have been developed for specialised computer hardware (for example, MPSrch running on the massively parallel MasPar supercomputer); however, these systems are expensive and rapidly become obsolete as hardware architectures develop and move on. Speed of execution is certainly an issue for database searching, and for both algorithms described so far, speed depends critically on the length of the query sequence and on the size of the database searched. The FastA and BLAST programs are essentially local similarity search methods that concentrate on finding short identical matches, which may contribute to a total match, using implementations that address issues of execution speed, without resorting to the use of specialised computer hardware.

6.13.1 FastA

The FastA algorithm, described by Lipman and Pearson in 1985, is based around the idea of identifying short words, or k-tuples, common to both sequences under comparison. K-tuple sizes of 1 or 2 residues are used in protein searches, while larger k-tuples (up to 6 bases) are used in DNA searches. Comparison of k-tuples, and their relative offsets between the two sequences, can be viewed as focusing on diagonal matches in a dynamic programming matrix. FastA uses a heuristic approach to join k-tuples that lie close together on the same diagonal. The regions formed in this way contain mismatches lying between matching k-tuples (FastA regions are analogous to segment pairs in BLAST, described below). If a significant

Pairwise alignment techniques

```
FASTA version 3.0t82 November 1, 1997
Please cite: W.R. Pearson & D.J. Lipman PNAS (1988) 85:2444-2448

>gi|631066|pir||JC2331 adrenergic receptor alpha 1A - human, 572 bases
 vs SWISS-PROT Protein Sequence Database (rel35) library
25083768 residues in 69113 sequences
 statistics extrapolated from 50000 to 68413 sequences
 Expectation_n fit: rho(ln(x))= 6.3487+/-0.000531; mu= 6.8138+/- 0.030;
 mean_var=205.1722+/-43.131, Z-trim: 515 B-trim: 2588 in 1/63

FASTA (3.08 July, 1997) function (optimized, blosum matrix) ktup: 2
 join: 37, opt: 25, gap-pen: -12/ -2, width: 16 reg.-scaled
 Scan time: 12.420
The best scores are:                              initn init1 opt z-sc E(68413)
SW:A1AA_HUMAN P25100 homo sapiens (human ( 572) 3836 3836 3836 2695.2 1.8e-143
SW:A1AA_RAT P23944 rattus norvegicus (ra ( 561) 2691 2259 3156 2220.5 4.9e-117
SW:A1AB_RAT P15823 rattus norvegicus (ra ( 515) 1618 1019 1617 1146.5 3.2e-57
SW:A1AB_HUMAN P35368 homo sapiens (human ( 519) 1620 1011 1615 1145.0 3.9e-57
SW:A1AB_MESAU P18841 mesocricetus auratu ( 515) 1618 1019 1608 1140.2 7.3e-57
SW:A1AC_HUMAN P35348 homo sapiens (human ( 466) 1423  935 1464 1040.1 2.7e-51
SW:A1AC_RAT P43140 rattus norvegicus (ra ( 466) 1439  933 1458 1035.9 4.7e-51
SW:A1AC_BOVIN P18130 bos taurus (bovine) ( 466) 1417  922 1443 1025.4 1.8e-50
SW:A1AA_ORYLA Q91175 oryzias latipes (me ( 470) 1413  956 1434 1019.1 4e-50
SW:A1AB_CANFA P11615 canis familiaris (d ( 417) 1372  772 1366 972.2 1.7e-47

-----

>>SW:A1AA_RAT P23944 rattus norvegicus (rat). alpha-1a a (561 aa)
 initn: 2691 init1: 2259 opt: 3156 Z-score: 2220.5 expect() 4.9e-117
Smith-Waterman score: 3156; 85.315% identity in 572 aa overlap

            10        20        30        40        50        60
gi|631 MTFRDLLSVSFEGPRPDSSAGGSSAGGGGGGAGGAAPSEGPAVGGVPGGAGGGGGVVGAG
       :::::::.:::.:::.: : :: :::::: : :::::.:
SW:A1A MTFRDILSVTFEGPRSSSSTGGSGAGGGAGTVG---P-EGGAVGGVPG-ATGGGAVVGTG
            10        20        30            40        50

            70        80        90       100       110       120
gi|631 SGEDNRSSAGEPGSAGAGGDVNGTAAVGGLVVSAQGVGVGVFLAAFILMAVAGNLLVILS
       ::::::.::.::::.: :.:.:::::::::::::::::::::::::::.::::::::::::
SW:A1A SGEDNQSSTGEPGAA-ASGEVNGSAAVGGLVVSAQGVGVGVFLAAFILTAVAGNLLVILS
           60        70        80        90       100       110

           130       140       150       160       170       180
gi|631 VACNRHLQTVTNYFIVNLAVADLLLSATVLPFSATMEVLGFWAFGRAFCDVWAAVDVLCC
       ::::::::::::::::::::::::::::.:::::::::::::::::::.:::::::::::::
SW:A1A VACNRHLQTVTNYFIVNLAVADLLLSAAVLPFSATMEVLGFWAFGRTFCDVWAAVDVLCC
           120       130       140       150       160       170

           190       200       210       220       230       240
gi|631 TASILSLCTISVDRYVGVRHSLKYPAIMTERKAAAILALLWVVALVVSVGPLLGWKEPVP
       :::::::::::::::::::::::::::::::::::::::::.::::::::::::::::::::
SW:A1A TASILSLCTISVDRYVGVRHSLKYPAIMTERKAAAILALLWAVALVVSVGPLLGWKEPVP
           180       190       200       210       220       230

-----

           490       500       510       520       530       540
gi|631 QAPVASRRKPPSAFREWRLLGPFRRPTTQLRAKVSSLSHKIPAGGAQRAEAACAQRSEVE
       :  :.: ::: ::.::::::::.::::::::::::::::::: .: :.:::.:: :::::
SW:A1A QDSVSSSRKPASALREWRLLGPLQRPTTQLRAKVSSLSHKIRSG-ARRAETACALRSEVE
           480       490       500       510       520

           550       560       570
gi|631 AVSLGVPHEVAEGATCQAYELADYSNLRETDI
       ::::.::.. ::.. ::::: .::::::::::
SW:A1A AVSLNVPQDGAEAVICQAYEPGDYSNLRETDI
           530       540       550       560

-----
```

Figure 6.8 Excerpt from a typical FastA output (----- denotes excised material).

to compute gapped alignments that incorporate the ungapped regions.

A typical output from a FastA search is presented in Figure 6.8. The name and version of the program are given at the top of the file, with the appropriate citation to use in any published results arising from use of the program. The query sequence is indicated (here, human alpha-1A adrenergic receptor), together with the name and version of the search database (in this case, SWISS-PROT release 35). This information is followed by a print-out of a range of parameters used by the algorithm and the run-time of the program (12.42 seconds).

Following the program statistics come the results of the database search itself, commencing with a list of a user-defined number of matches with the query sequence (for convenience, only 10 hits are shown, but normal searches would seek at least the top 50). This list contains a source database identifier (here, for example, SW denotes SWISS-PROT), and the database ID code, accession number and title of the matched sequences. The length (in amino acid residues) of each of the retrieved matches is indicated in brackets. This information continues with various initial and optimised scores calculated by the program, and, most importantly, an Expect- or E-value (see Section 6.7.3) that allows the user to assess the likelihood of the match being true (an E-value approaching zero indicates that virtually no matches with a similar score would be expected simply by chance). In this example, the top 10 hits all have very low E-values, indicating that they are all likely to be real.

After the search summary, the output presents the complete pairwise alignments of a user-defined number of hits with the query sequence (users may request to see more or fewer alignments than are specified in the search summary, or may opt not to view any alignments at all – the tree-friendly option!). Within the alignments, identities are indicated by the ':' character, and similarities by a '.' (inspection of the program parameters at the top of the file indicates that the scoring matrix used here was one of the BLOSUM series). Rulers above and below the alignments denote absolute residue numbers for the sequences (i.e., ignoring gaps).

6.13.2 BLAST

The BLAST (Basic Local Alignment Search Tool) algorithm was described by Altschul *et al.* in 1990. It has become popular largely because implementations of it have been very efficient, and it has been optimised to work with parallel UNIX architectures from an early stage. These characteristics result in rapid turnaround of sequence searches from public servers.

The algorithm itself is straightforward, the important concept being that of the segment pair. Given two sequences, a segment pair is defined as a pair of sub-sequences of the same length that form an ungapped alignment. BLAST calculates all segment pairs between the query and the database sequences, above a scoring threshold. The algorithm searches for fixed-length hits, which are then extended until certain threshold parameters are achieved. The resulting high-scoring pairs (HSPs) form the basis of the ungapped alignments that characterise BLAST output; four HSPs are illustrated in the example output shown in Figure 6.9.

The name and version of the program are given at the top of the file, with the appropriate citation to use in any published results arising from use of the program. The query sequence is indicated (again, human alpha-1A adrenergic receptor), together with the name of the search database (in this case, a non-redundant version of SWISS-PROT).

Following the header information come the results of the database search, commencing with a list of a user-defined number of matches with the query sequence (again, only 10 hits are shown, but normal searches would seek at least 50). This list contains a source database identifier (here, sp denotes SWISS-PROT), and the database accession number, ID code and title of the matched sequences. The information continues with the highest score of the set of matching segment pairs for the given sequence (the number of HSPs is given by the parameter N). Importantly, there then follows a probability- or p-value (see Section 6.7.3) that allows the user to assess the likelihood of the match being true (a p-value approaching zero indicates that the probability of this match having arisen by chance is virtually zero). In this example, the top 10 hits all have very low p-values, indicating that they are all likely to be real.

After the search summary, the output presents the ungapped pairwise alignments of the HSPs for each of a user-defined number of hits with the query sequence (as with FastA, users may request to see more or fewer alignments than are specified in the search summary, or may take the forest-friendly option not to view the alignments at all). For each aligned HSP, the beginning and end locations within the sequence are marked, and identities between them are indicated by the corresponding amino acid symbol. In this example, the program SEG has been implemented as part of the BLAST routine. The purpose of SEG is to mask out so-called low-complexity regions (i.e., regions within a sequence that have high densities of particular residues, e.g. GAPGAPGAPGAP... such as occurs in repetitive, often tightly structured sequences such as collagen), which would otherwise result in large numbers of spurious high-scoring matches swamping the hitlist. Regions masked by SEG are denoted by strings of Xs in the query sequence.

Gapped BLAST

The original BLAST program, described above, suffers from the limitation that it can only produce ungapped alignments. Experience shows that often several ungapped non-overlapping alignments result from a match to a single database sequence (see, for example, the number of HSPs given by the N parameter in Figure 6.9, which for these top 10 hits is greater than 1). Intuitively, we know that these alignments may be linked together into a larger (and perhaps biologically more realistic) alignment, in which the reliable HSPs are sewn together by less reliable gapped regions.

To address this shortcoming, a modification of the algorithm has been introduced for generating gapped alignments (Altshul *et al.*, 1997). The new algorithm seeks only one, rather than all, ungapped alignments that make up a significant match, and hence speeds the initial database search.

```
BLASTP 1.4.11 [24-Nov-97] [Build 24-Nov-97]

Reference: Altschul, Stephen F., Warren Gish, Webb Miller, Eugene W. Myers, and David
J. Lipman (1990). Basic local alignment search tool. J. Mol. Biol. 215:403-10.
Query= gi|631066|pir||JC2331 adrenergic receptor alpha 1A - human (572 letters)
Database: Non-redundant SwissProt (74,037 sequences; 26,661,674 total letters)
Searching....................................................done

                                                           Smallest
                                                             Sum
                                                    High  Probability
Sequences producing High-scoring Segment Pairs:     Score   P(N)      N

sp|P25100|A1AD_HUMAN ALPHA-1D ADRENERGIC RECEPTOR (ALPHA ... 1513  5.5e-266  4
sp|O02666|A1AD_RABIT ALPHA-1D ADRENERGIC RECEPTOR (ALPHA ... 1465  3.9e-242  4
sp|P23944|A1AD_RAT   ALPHA-1D ADRENERGIC RECEPTOR (ALPHA ... 1416  2.0e-228  5
sp|P97714|A1AD_MOUSE ALPHA-1D ADRENERGIC RECEPTOR (ALPHA ... 1411  5.1e-220  3
sp|P15823|A1AB_RAT   ALPHA-1B ADRENERGIC RECEPTOR (ALPHA ...  650  9.2e-130  2
sp|P18841|A1AB_MESAU ALPHA-1B ADRENERGIC RECEPTOR (ALPHA ...  650  9.2e-130  2
sp|P35368|A1AB_HUMAN ALPHA-1B ADRENERGIC RECEPTOR (ALPHA ...  643  8.8e-129  2
sp|P97717|A1AB_MOUSE ALPHA-1B ADRENERGIC RECEPTOR (ALPHA ...  629  8.2e-127  2
sp|P35348|A1AA_HUMAN ALPHA-1A ADRENERGIC RECEPTOR (ALPHA ...  589  4.2e-118  2
sp|O02824|A1AA_RABIT ALPHA-1A ADRENERGIC RECEPTOR (ALPHA ...  591  1.1e-117  2

-----

sp|P25100|A1AD_HUMAN ALPHA-1D ADRENERGIC RECEPTOR (ALPHA 1D-ADRENOCEPTOR) Length = 572

 Score = 89 (41.7 bits), Expect = 5.5e-266, Sum P(4) = 5.5e-266
 Identities = 17/17 (100%), Positives = 17/17 (100%)

Query:     1 MTFRDLLSVSFEGPRPD 17
             MTFRDLLSVSFEGPRPD
Sbjct:     1 MTFRDLLSVSFEGPRPD 17

 Score = 1513 (708.4 bits), Expect = 5.5e-266, Sum P(4) = 5.5e-266
 Identities = 299/348 (85%), Positives = 299/348 (85%)

Query:    63 EDNRXXXXXXXXXXXXXXXXDVNGTAAVGGLVVSAQGVGVGVFLAAFILMAVAGNLLVILSVA 122
             EDNR            DVNGTAAVGGLVVSAQGVGVGVFLAAFILMAVAGNLLVILSVA
Sbjct:    63 EDNRSSAGEPGSAGAGGDVNGTAAVGGLVVSAQGVGVGVFLAAFILMAVAGNLLVILSVA 122

Query:   123 CNRHLQTVTNYFIVNLAVADLLLSATVLPFSATMEVLGFWAFGRAFCDVWAAVDVLCCTA 182
             CNRHLQTVTNYFIVNLAVADLLLSATVLPFSATMEVLGFWAFGRAFCDVWAAVDVLCCTA
Sbjct:   123 CNRHLQTVTNYFIVNLAVADLLLSATVLPFSATMEVLGFWAFGRAFCDVWAAVDVLCCTA 182

Query:   183 SILSLCTISVDRYVGVRHSLKYPAIMTERKXXXXXXXXXXXXXXXXXXXXXXXXXGWKEPVPPD 242
             SILSLCTISVDRYVGVRHSLKYPAIMTERK          GWKEPVPPD
Sbjct:   183 SILSLCTISVDRYVGVRHSLKYPAIMTERKAAAILALLWVVALVVSVGPLLGWKEPVPPD 242

Query:   243 ERFCGITEEAGYAVFSSVCSFYLPMXXXXXXXXXXXXXXXXXSTTRSLEAGVKRERGKASEV 302
             ERFCGITEEAGYAVFSSVCSFYLPM          STTRSLEAGVKRERGKASEV
Sbjct:   243 ERFCGITEEAGYAVFSSVCSFYLPMAVIVVMYCRVYVVARSTTRSLEAGVKRERGKASEV 302

Query:   303 VLRIHCRGAATGADGAHGMRSAKGHTFRSSLSVRLLKFSREKKAAKTLAIVVGVFVLCWF 362
             VLRIHCRGAATGADGAHGMRSAKGHTFRSSLSVRLLKFSREKKAAKTLAIVVGVFVLCWF
Sbjct:   303 VLRIHCRGAATGADGAHGMRSAKGHTFRSSLSVRLLKFSREKKAAKTLAIVVGVFVLCWF 362

Query:   363 PFFFVLPLGSLFPQLKPSEGVFKVIFWLGYFNSCVNPLIYPCSSREFK 410
             PFFFVLPLGSLFPQLKPSEGVFKVIFWLGYFNSCVNPLIYPCSSREFK
Sbjct:   363 PFFFVLPLGSLFPQLKPSEGVFKVIFWLGYFNSCVNPLIYPCSSREFK 410

 Score = 101 (47.3 bits), Expect = 5.5e-266, Sum P(4) = 5.5e-266
 Identities = 17/17 (100%), Positives = 17/17 (100%)

Query:   433 VYGHHWRASTSGLRQDC 449
             VYGHHWRASTSGLRQDC
Sbjct:   433 VYGHHWRASTSGLRQDC 449

 Score = 387 (181.2 bits), Expect = 5.5e-266, Sum P(4) = 5.5e-266
 Identities = 78/93 (83%), Positives = 78/93 (83%)

Query:   480 MQAPVASRRKPPSAFREWRLLGPFRRPTTQLRAKVSSLSHKIPXXXXXXXXXXXXXXXSEV 539
             MQAPVASRRKPPSAFREWRLLGPFRRPTTQLRAKVSSLSHKI          SEV
Sbjct:   480 MQAPVASRRKPPSAFREWRLLGPFRRPTTQLRAKVSSLSHKIRAGGAQRAEAACAQRSEV 539

Query:   540 EAVSLGVPHEVAEGATCQAYELADYSNLRETDI 572
             EAVSLGVPHEVAEGATCQAYELADYSNLRETDI
Sbjct:   540 EAVSLGVPHEVAEGATCQAYELADYSNLRETDI 572

-----
```

Figure 6.9 Excerpt from a typical BLAST output (----- denotes excised material).

Dynamic programming is used to extend a central pair of aligned residues in both directions to yield the final gapped alignment. Having dropped the requirement to find all ungapped alignments independently, the new algorithm is three times faster than its predecessor. Further extensions to BLAST, which effectively provide a hybrid of pairwise and multiple alignment methods, are discussed in the following chapter.

6.14 Summary

- Database interrogation can take the form of text queries or sequence similarity searches. To identify an evolutionary relationship between a newly determined sequence and a known gene family, the extent of shared similarity must be assessed.

- An algorithm is a set of steps that define a computational process; a program is an implementation of an algorithm. There may be different implementations of the same algorithm, which ought to (but may not!) give the same results.

- The simplest way to compare two sequences is to align them by inserting gap characters to bring them into vertical register. Counting the matched character positions gives a naïve alignment score.

- Computing alignment scores is more exacting for long, dissimilar sequences with disparate lengths. Scoring penalties are employed to minimise the number and length of gaps, and matrices are used to score both identical and similar residues.

- Identity matrices are sparse (most matrix elements score zero) and so have poor diagnostic power. Similarity matrices weight non-identical residue matches according to observed substitution rates across large evolutionary distances. Such matrices are noisy because they boost both random matches and weak signals. Distinguishing low-scoring biological signals from high-scoring noise is a central challenge of sequence analysis.

- Scores in the Dayhoff Mutation Data Matrix are based on the concept of the Point Accepted Mutation (PAM). An evolutionary distance of 250 PAMs gives similarity scores equivalent to 20% matches remaining between two sequences. This is the Twilight Zone, hence PAM 250 is often used as the default matrix in comparison programs.

- Scores in the BLOSUM matrices are derived from observed substitutions in blocks of aligned sequences from the BLOCKS database. They were designed to detect distant similarities more reliably than the Dayhoff matrices, which can only *infer* distant relationships because their substitution rates were derived from highly similar sequences.

- That a program may align two sequences is not proof that a relationship exists between them. Statistical values are used to indicate the level of confidence that should be attached to an alignment; for pairwise alignments, these are usually formulated as probability (p) values or expected frequency (E) values.

- A basic method of comparing two sequences is the dotplot. This is a graph in which the sequences lie on the *x*- and *y*-axes, and crosses/dots are plotted at all positions where identical residues are observed. For identical sequences, this leads to an unbroken diagonal line across the plot, while similar sequences give rise to broken diagonals.

- Alignments are models that reflect different biological perspectives. One model is therefore no more right or wrong than another. Two general approaches consider similarity (a) across the full extent of sequences (global alignment – the Needleman and Wunsch algorithm) and (b) across only parts of the sequences (local alignment – the Smith–Waterman algorithm).

- The Needleman and Wunsch and Smith–Waterman algorithms exploit dynamic programming, whereby a solution to a problem is built by solving smaller, tractable sub-problems. The optimal alignment is chosen from a set of high-scoring alternatives. Such methods are prohibitively time-consuming for large numbers of sequences.

- The FastA and BLAST programs are local similarity search methods that concentrate on finding short identical matches, which may contribute to a total match. Speed issues are addressed using heuristics. A recent, faster implementation of BLAST is able to generate gapped alignments.

6.15 Further reading

Alignment methods

DAYHOFF, M.O., SCHWARTZ, R.M. and ORCUTT, B.C. (1978) In *Atlas of Protein Sequence and Structure*, Vol. 5, Suppl. 3, Dayhoff, M.O. (ed.), NBRF, Washington, DC, p. 345.

HENIKOFF, S. and HENIKOFF, J.G. (1992) Amino acid substitution matrices from protein blocks. *Proceedings of the National Academy of Sciences, USA*, **89**, 10915–10919.

NEEDLEMAN, S.B. and WUNSCH, C.D. (1970) A general method applicable to the search for similarities in the amino acid sequences of two proteins. *Journal of Molecular Biology*, **48**, 443–453.

SMITH, T.F. and WATERMAN, M.S. (1981) Identification of common molecular subsequences. *Journal of Molecular Biology*, **147**, 195–197.

Database search methods

ALTSCHUL, S.F., GISH, W., MILLER, W., MYERS, E.W. and LIPMAN, D.J. (1990) Basic local alignment search tool. *Journal of Molecular Biology*, **215**, 403–410.

ALTSCHUL, S.F., BOGUSKI, M.S., GISH, W. and WOOTTON, J.C. (1994) Issues in searching molecular sequence databases. *Nature Genetics*, **6**, 119–129.

ALTSCHUL, S.F., MADDEN, T.L., SCHÄFFER, A.A., ZHANG, J., ZHANG, Z., MILLER, W. and LIPMAN, D.J. (1997) Gapped BLAST and PSI-BLAST: a new generation of protein database search programs. *Nucleic Acids Research*, **25**(17), 2289-3402.

LIPMAN, D.J. and PEARSON, W.R. (1985) Rapid and sensitive protein similarity searches. *Science*, **227**, 1435–1441.

CHAPTER SEVEN

Multiple sequence alignment

7.1 Introduction

Pairwise comparison is fundamental to sequence analysis. However, analysis of groups of sequences that form **gene families** requires the ability to make connections between more than two members of the group, in order to reveal subtle conserved family characteristics. The process of multiple alignment can be regarded as an exercise in enhancing the signal-to-noise ratio within a set of sequences, which ultimately facilitates the elucidation of biologically significant motifs. In this chapter, we review a range of different approaches to multiple alignment, from fully manual methods to widely used automatic techniques.

7.2 The goal of multiple sequence alignment

The goal of multiple sequence alignment is to generate a concise, information-rich summary of sequence data in order to inform decision-making on the relatedness of sequences to a gene family. Sometimes, indeed, multiple alignments may be used to express the *dissimilarity* between a set of sequences. Alignments should be regarded as models (in both a mathematical and a biological sense) that can be used to test hypotheses. Just as we saw that there is nothing inherently correct, or incorrect, about any particular pairwise alignment, we see here that the same maxim holds for multiple alignments. The question always to ask of an alignment is 'Does this model accurately reflect the known biological evidence?'

In order to address this question, the sequence analyst needs a number of tools; in addition to tried-and-tested automatic programs, arguably the most important of these is a good manual multiple sequence editor.

It should be appreciated that alignment models may be generated from many different starting points, reflecting not only the choice of alignment tool (e.g., whether manual or automatic), but also, just as importantly, the underlying biological basis for the comparison. Essentially, there are two main perspectives on the construction of alignments: the first approach is guided by the comparison of similar strings of amino acid residues (for example, taking into account physicochemical properties, mutability data, etc.); the second results from comparison at the level of secondary or tertiary structure, where alignment positions are determined solely on the basis of structural equivalence. Alignments generated from these rather different viewpoints often show significant disparities, begging the all-too-familiar question, 'Which is correct?' The answer is that both are equally valid; each is a model that reflects a specific view of the known biology.

Both sequence- and structure-based alignments are imperfect models, as neither can reflect all of the evidence. This should not be surprising, as protein sequences are closer to DNA, where on-going genetic events actually occur, whereas protein structure reflects a biological state after post-translational modification, and other molecular interactions, have lent a protein its stable native fold. Criticisms of purely sequence-based alignments are usually framed by comparison with corresponding structural alignments, which are often assumed to be 'correct'. Obviously, if structural data are available, they may be helpful in guiding the analysis. However, the more usual situation is that no reliable structural data exist (i.e., from a physical experiment). Here, the analyst must rely on sequence similarity, together with biochemical evidence, in order to build a satisfactory multiple sequence model.

7.3 Multiple sequence alignment: a definition

A multiple sequence alignment is a 2D table, in which the rows represent individual sequences, and the columns the residue positions. Sequences are laid onto this grid in such a manner that (a) the relative positioning of residues within any one sequence is preserved, and (b) similar residues in all the sequences are brought into vertical register (see Table 7.1). We call the residue position in an unaligned sequence, the absolute position (for example, if residue 3 in sequence I is Glycine, then absolute position I3 is

Table 7.1 Definition of multiple sequence alignment. Here, a small alignment of five short sequences (I–V) is presented. The sequences have been arranged so that the most similar residues are brought into vertical register, through the use of gaps ('-' characters), while the order of residues in each sequence is preserved.

	1	2	3	4	5	6	7	8	9	10
I	Y	D	G	G	A	V	–	E	A	L
II	Y	D	G	G	–	–	–	E	A	L
III	F	E	G	G	I	L	V	E	A	L
IV	F	D	–	G	I	L	V	Q	A	V
V	Y	E	G	G	A	V	V	Q	A	L

always Glycine – a chemical bond would need to be broken to change it). By contrast, we call the aligned residue position, the relative position. All residues in any single column of an alignment will have the same relative position and, almost certainly, different absolute positions (unless all the sequences are identical!). This may seem perverse, but all that needs to be considered is that the absolute position is a property of the sequence, while the relative position is a property of the alignment.

7.4 The consensus

The alignment table can be summarised in a single line – a pseudo-sequence – normally added at the end of the alignment. This pseudo-sequence consists of symbols that summarise the character of the alignment at each vertical position, or column, as shown in Table 7.2.

Table 7.2 Multiple alignment and the consensus sequence. The consensus is shown on the last line of the alignment table. Here, the consensus has been calculated using the following rules: if only one residue symbol is present, use an uppercase letter in the consensus; if the majority of symbols are one letter, use a lowercase letter; if equal numbers of different residues are present, show all residues in the consensus.

	1	2	3	4	5	6	7	8	9	10
I	Y	D	G	G	A	V	–	E	A	L
II	Y	D	G	G	–	–	–	E	A	L
III	F	E	G	G	I	L	V	E	A	L
IV	F	D	–	G	I	L	V	Q	A	V
V	Y	E	G	G	A	V	V	Q	A	L
	y	d	G	G	A/I	V/L	V	e	A	l

The consensus need not take the form of a single sequence line, however. Some methods use a weight matrix approach to summarise the whole alignment (e.g., profiles – see below). Other methods automatically seek out conserved, ungapped blocks of residues within alignments, which are then converted to position-specific scoring matrices (e.g., blocks). Finally, highly specific, relatively short ungapped motifs are manually extracted from alignments and used to generate unweighted scoring matrices (e.g., fingerprints). Profiles, blocks and fingerprints are discussed in detail in Chapters 3 and 8.

7.5 Computational complexity

Pairwise alignment techniques generally use processing time and memory space related to the product of the lengths of the sequences being compared ($O(m_1 m_2)$, where O is known as the order of the time taken by the algorithm, and m_1 and m_2 are the sequence lengths). By extending pairwise comparison to three dimensions (by adding another axis to the 2D scoring matrix) we have a time complexity of $O(m_1 m_2 m_3)$, where m_3 is the length of the third sequence.

When considering more sequences, the time complexity becomes $O(m_1 m_2 m_3 ... m_l)$, where m_l is the length of the last sequence in the comparison set, or more concisely $O(m^n)$, where n is the number of sequences and m is the length of the sequences. Thus, the time taken to compute an alignment rises exponentially the more sequences there are to be aligned.

Various methods have been developed that use heuristics to reduce the time to find good (not necessarily optimal) alignments (e.g., using algorithms that exploit sub-sequences, trees, consensus sequences, clustering and templates). Some approaches combine dynamic programming with heuristics. Such techniques include aligning all pairs of sequences, aligning each sequence with one specific sequence, aligning sequences in arbitrary order, or aligning sequences following the branching order of a phylogenetic tree. The results of such approaches tend not to be optimal, and normally require at least $n - 1$ pairwise alignments, where n is the number of sequences in the set to be aligned.

7.6 Manual methods

Manual methods tend to be glossed over in the literature because the results of human tinkering with sequences are regarded as being subjective. However, we should not be blind to the bias introduced into scientific analysis of a set of sequences by relying entirely on one computational method. Ultimately, to combine the results of tried-and-trusted programs with biological evidence gleaned from experiment or from examination of the literature, does necessitate manual intervention.

Numerous sequence analysis packages have been developed, some of which offer manual alignment editors (the best offer a choice of manual editing facilities coupled with popular automatic algorithms). To facilitate interactive alignment, colour is often used to assist the eye in spotting similarities. Many editors use non-intuitive colouring schemes to depict amino acid properties, resulting in the loss of much of the useful information sequestered in the alignment. Appropriate choice of colour, however, allows user-defined properties to be depicted in an immediately informative way, no matter how large the alignment; and it offers a rapid and informed means of selecting residues suitable for mutagenesis studies by revealing regions crucial to the structure or function of a protein (e.g., critically conserved residues, or residue groups, can be seen at a glance; unusual mutations may stand proud against a smooth backdrop of conservation; and mutational hotspots can be readily pinpointed). Although the actual choice of colour is ultimately a subjective one, schemes that are consistent with standard physical modelling components and 3D graphics packages tend to be the most helpful. An example colour scheme is given in Table 7.3.

Another important feature of alignment programs is the ability to derive a quantitative evaluation of the relatedness of all sequence pairs (e.g., through calculation of residue identities, or of combined identities and similarities). This allows an at-a-glance summary of the evolutionary distances

Table 7.3 Typical amino acid property groupings and example alignment colouring scheme, broadly consistent with modelling components and 3D graphics packages.

Residue	Property	Colour
Asp, Glu	Acidic	red
His, Arg, Lys	Basic	blue
Ser, Thr, Asn, Gln	Polar neutral	green
Ala, Val, Leu, Ile, Met	Hydrophobic aliphatic	white
Phe, Try, Trp	Hydrophobic aromatic	purple
Pro, Gly	Special structural properties	brown
Cys	Disulphide bond former	yellow

between aligned sequences, and may help to assess alignment quality (e.g., if similarity values appear unexpectedly low, either the sequences are truly distant relatives, or the alignment contains errors).

Point-and-click windowing interfaces are common in recent packages, where, for example, the alignment editor sits at the centre of a range of sequence analysis tools (such packages are discussed in more detail in Chapter 10).

As we will see, in Section 7.9, alignment quality from automatic methods tends to be greatest when similar sequences of similar length are being compared, and poorest when aligning distantly related sequences with disparate lengths. Manual alignment editors are thus essential tools for polishing the output from automatic programs. Besides, it is always good working practice to use different methods for sequence analysis, and to combine results of the different approaches. As a learning aid, the student should therefore take every opportunity to work with sequences in a good manual alignment system and compare results with several automated methods.

7.7 Simultaneous methods

The essence of simultaneous methods is to align all the sequences in a given set at once, rather than taking a progressive approach and aligning pairs of sequences, or building sequence clusters. The basic idea is an extension of the 2D dynamic programming matrix, presented in Chapter 6, to three or more dimensions; the number of dimensions in the matrix reflects the number of sequences to be aligned. These methods tend to have exacting computer system resource requirements and, as a result, work best on small sets of short sequences.

7.8 Progressive methods

In this category, probably the best known program is Clustal, which has its roots in the 1987 method of Feng and Doolittle. As we have seen, using a multi-dimensional dynamic programming matrix is not practical for calculating alignments on realistic sets of data. Most programs use heuristics to arrive at an alignment in a timely and cost-efficient manner. The Clustal approach exploits the fact that similar sequences are likely to be evolutionarily related. Thus, the method aligns sequences in pairs, following the branching order of a family tree. Similar sequences are aligned first, and more distantly related sequences are added later. Once pairwise alignment scores for each sequence relative to all others have been calculated, they are used to cluster the sequences into groups, which are then aligned against each other to generate the final multiple alignment. As part of its operation, the program can produce information required to produce a phylogenetic tree.

Clustal has been through various revisions; ClustalW, described by Thompson *et al.* in 1994, uses the positioning of gaps in closely related sequences to guide the insertion of gaps into those that are more distant. Similarly, information compiled during the alignment process about the variability of the most similar sequences is used to help vary the gap penalties on a residue and position-specific basis.

Since Clustal is widely and freely available, it is frequently used in sequence analysis, together with a variety of other tools. Thus, Clustal accepts alignments in several formats: EMBL/SWISS-PROT, NBRF/PIR, Pearson/FastA, GCG/MSF, and Clustal's own format. Output may be requested in Clustal format or in formats compatible with the GDE, Phylip or GCG packages.

Part of a typical alignment of a set of adrenergic G-protein-coupled receptors (GPCRs) is illustrated in Figure 7.1; the alignment was generated using default parameters, but toggling the output to GCG/MSF format, as shown. The sequences are clearly highly similar and of similar lengths, resulting in a high-quality alignment. From the region depicted, it is evident that there are two short gapped regions: in the two A2AA receptors, each has a three-residue insertion (GPQ and GQQ) relative to the other sequences, hence the region of gaps at the end of the first paragraph; and in the two A2AB receptors, each has a single residue insertion, resulting in the gaps at the end of the second paragraph.

7.9 Databases of multiple alignments

The power of multiple sequence analysis lies in the ability to draw together related sequences from various species and express the degree of similarity in a relatively concise format. The wealth of information presented in alignments can also be used to enhance the sensitivity of database searches (see Section 7.10); hence, demand for readily available, high-quality alignments has led to the production of multiple alignment databases.

Today, there are numerous alignment databases accessible via the Web: some of these result from automated approaches to cluster the primary sequence resources into families; others result from endeavours to produce gene family discriminators for inclusion in secondary databases, via manual

```
A1AA_HUMAN SLKYPAIMTE RKAAAILALL WVVALVVSVG PLLGWKEPVP P.....DERF
A1AA_RAT   SLKYPAIMTE RKAAAILALL WAVALVVSVG PLLGWKEPVP P.....DERF
A1AB_HUMAN SLQYPTLVTR RKAILALLSV WVLSTVISIG PLLGWKEPAP N.....DDKE
A1AB_RAT   SLQYPTLVTR RKAILALLSV WVLSTVISIG PLLGWKEPAP N.....DDKE
A1AC_HUMAN PLRYPTIVTQ RRGLMALLCV WALSLVISIG PLFGWRQPAP E.....DETI
A1AC_RAT   PLRYPTIVTQ RRGVRALLCV WVLSLVISIG PLFGWRQPAP E.....DETI
A2AA_HUMAN AIEYNLKRTP RRIKAIIITV WVISAVISFP PLISIEKKGG GGGPQPAEPR
A2AA_RAT   AIEYNLKRTP RRIKAIIVTV WVISAVISFP PLISIEKKGA GGGQQPAEPS
A2AB_HUMAN ALEYNSKRTP RRIKCIILTV WLIAAVISLP PLIYKGDQGP QP...RGRPQ
A2AB_RAT   ALEYNSKRTP CRIKCIILTV WLIAAVISLP PLIYKGDQRP DA...RGLPQ
A2AC_HUMAN AVEYNLKRTP RRVKATIVAV WLISAVISFP PLVSLYRQPD G....AAYPQ
A2AC_RAT   AVEYNLKRTP RRVKATIVAV WLISAVISFP PLVSFYRRPD G....AAYPQ
A2AD_HUMAN AVEYNLKRTP RRVKATIVAV WLISAVISFP PLVSLYRQPD G....AAYPQ


A1AA_HUMAN CGITEEAGYA VFSSVCSFYL PMAVIVVMYC RVYVVARST. TRSLEAGVKR
A1AA_RAT   CGITEEVGYA IFSSVCSFYL PMAVIVVMYC RVYVVARST. TRSLEAGIKR
A1AB_HUMAN CGVTEEPFYA LFSSLGSFYI PLAVILVMYC RVYIVAKRT. TKNLEAGVMK
A1AB_RAT   CGVTEEPFYA LFSSLGSFYI PLAVILVMYC RVYIVAKRT. TKNLEAGVMK
A1AC_HUMAN CQINEEPGYV LFSALGSFYL PLAIILVMYC RVYVVAKRE. SRGLKSGLKT
A1AC_RAT   CQINEEPGYV LFSALGSFYV PLAIILVMYC RVYVVAKRE. SRGLKSGLKT
A2AA_HUMAN CEINDQKWYV ISSCIGSFFA PCLIMILVYV RIYQIAKRR. TRVPPSRRGP
A2AA_RAT   CKINDQKWYV ISSSIGSFFA PCLIMILVYV RIYQIAKRR. TRVPPSRRGP
A2AB_HUMAN CKLNQEAWYI LASSIGSFFA PCLIMILVYL RIYLIAKRSN RRGPRAKGGP
A2AB_RAT   CELNQEAWYI LASSIGSFFA PCLIMILVYL RIYVIAKRSH CRGLGAKRGS
A2AC_HUMAN CGLNDETWYI LSSCIGSFFA PCLIMGLVYA RIYRVAKRR. TRTLSEKRAP
A2AC_RAT   CGLNDETWYI LSSCIGSFFA PCLIMGLVYA RIYRVAKLR. TRTLSEKRGP
A2AD_HUMAN CGLNDETWYI LSSCIGSFFA PCLIMGLVYA RIYRVAKLR. TRTLSEKRAP
```

Figure 7.1 Typical multiple alignment output from ClustalW (in MSF format). The result shows part of an alignment of adrenergic receptors, the sequences of which are highly similar and of similar lengths. The output is read from left to right, in paragraphs of 50 residues, formatted in 10-residue chunks; gaps are denoted by the '.' character. Two small insertions are highlighted.

or automatic methods. It is beyond the scope of this book, and perhaps of little value, to attempt to describe all such resources; but it is perhaps instructive to reflect on the quality we might expect from manually and automatically derived alignments.

Let us return to the example of the adrenergic GPCRs discussed in Section 7.8. The GPCRs constitute a superfamily that, in addition to the adrenergic receptors, includes many other receptor families (e.g., dopaminergic, muscarinic, olfactory, gustatory, visual, cannabinoid, opioid, and many more). There are now more than 1000 GPCR sequences known, encompassing a large spread of the evolutionary tree; creating alignments of such large numbers of diverse sequences is a non-trivial task. Consider, then, an excerpt from an alignment provided by one of the automatically derived multiple alignment databases (Pfam), shown in Figure 7.2. The figure focuses on a portion of the alignment that includes several adrenergic receptors, together with receptors for dopamine and octopamine, and depicts precisely the same region as illustrated in Figure 7.1.

```
A1AA_HUMAN    SLKYP....AI..MTER.KAA..AILALL.WVV.AL.VVSVGP.LLG...WKEPV..PPDE....RF
A1AA_RAT      SLKYP....AI..MTER.KAA..AILALL.WAV.AL.VVSVGP.LLG...WKEPV..PPDE....RF
A1AB_CANFA    SLQYP....TL..VTRR.KAI..LALLGV.WVL.ST.VISIGP.LLG...WKEPA..PNDD....KE
A1AB_HUMAN    SLQYP....TL..VTRR.KAI..LALLSV.WVL.ST.VISIGP.LLG...WKEPA..PNDD....KE
A1AB_MESAU    SLQYP....TL..VTRR.KAI..LALLSV.WVL.ST.VISIGP.LLG...WKEPA..PNDD....KE
A1AB_RAT      SLQYP....TL..VTRR.KAI..LALLSV.WVL.ST.VISIGP.LLG...WKEPA..PNDD....KE
A1AC_BOVIN    PLRYP....TI..VTQK.RGL..MALLCV.WAL.SL.VISIGP.LFG...WRQPA..PEDE.....T
A1AC_HUMAN    PLRYP....TI..VTQR.RGL..MALLCV.WAL.SL.VISIGP.LFG...WRQPA..PEDE.....T
A1AC_RAT      PLRYP....TI..VTQR.RGV..RALLCV.WVL.SL.VISIGP.LFG...WRQPA..PEDE.....T
OAR_DROME     PINYA....QK..RTVG.RVL..LLISGV.WLL.SL.LISSPP.LIG...W.NDW..PDEFT..SAT
D1DR_CARAU    PFRYE....RK..MTPR.VAF..VMISGA.WTL.SV.LISFIPVQLK...WHKAQ..PIGFL..EVN
D1DR_FUGRU    PFRYE....RK..MTPK.VAC..LMISVA.WTL.SV.LISFIPVQLN...WHKAQ..TASYVELNGT
DADR_DIDMA    PFRYE....RK..MTPK.AAF..ILISVA.WTL.SV.LISFIPVQLN...WHKARPLSSPDG..NVS
.....                .....


A1AA_HUMAN    ...............CGI..TE....................EAG...YA.....VF......SS
A1AA_RAT      ...............CGI..TE....................EVG...YA.....IF......SS
A1AB_CANFA    ...............CGV..TE....................EPF...YA.....LF......SS
A1AB_HUMAN    ...............CGV..TE....................EPF...YA.....LF......SS
A1AB_MESAU    ...............CGV..TE....................EPF...YA.....LF......SS
A1AB_RAT      ...............CGV..TE....................EPF...YA.....LF......SS
A1AC_BOVIN    ..............ICQI..NE....................EPG...YV.....LF......SA
A1AC_HUMAN    ..............ICQI..NE....................EPG...YV.....LF......SA
A1AC_RAT      ..............ICQI..NE....................EPG...YV.....LF......SA
OAR_DROME     ..............PCEL..TS....................QRG...YV.....IY......SS
D1DR_CARAU    ..............ASRR...DLPTDNC........DSSL....NRT...YA.....IS......SS
D1DR_FUGRU    ..............YAGD..LPPDNCD.........SSL....NRT...YA.....IS......SS
DADR_DIDMA    ..............SQDE...TMDNCD.........SSL....SRT...YA.....IS......SS
.....                .....


A1AA_HUMAN    V.CSFY...LPMAVIV.VMY.CRV.YVV....ARS..TTRSLEA.GVKR
A1AA_RAT      V.CSFY...LPMAVIV.VMY.CRV.YVV....ARS..TTRSL.....EA
A1AB_CANFA    L.GSFY...IPLAVIL.VMY.CRV.YIV....AKR..TTKNL.....EA
A1AB_HUMAN    L.GSFY...IPLAVIL.VMY.CRV.YIV....AKR..TTKNL.....EA
A1AB_MESAU    L.GSFY...IPLAVIL.VMY.CRV.YIV....AKR..TTKNL.....EA
A1AB_RAT      L.GSFY...IPLAVIL.VMY.CRV.YIV....AKR..TTKNL.....EA
A1AC_BOVIN    L.GSFY...VPLTIIL.VMY.CRV.YVV....AKR..ESRGLKS.GLKT
A1AC_HUMAN    L.GSFY...LPLAIIL.VMY.CRV.YVV....AKR..ESRGLKS.GLKT
A1AC_RAT      L.GSFY...VPLAIIL.VMY.CRV.YVV....AKR..ESRGLKS.GLKT
OAR_DROME     L.GSFF...IPLAIMT.IVY.IEI.FVA....TRR..RLRERA..RANK
D1DR_CARAU    L.ISFY...IPVAIMI.VTY.TQI.YRI....AQK..QIRRIS..ALER
D1DR_FUGRU    L.ISFY...IPVAIMI.VTY.TRI.YRI....AQK..QIRRIS..ALER
DADR_DIDMA    L.ISFY...IPVAIMI.VTY.TRI.YRI....AQK..QIRRIS..ALER
.....                .....
```

Figure 7.2 Excerpt from an alignment of 597 G-protein-coupled receptors (GPCRs) taken from the Pfam database. The result depicts the same part of the alignment as shown in Figure 7.1, but the alignment has been extended to include other members of the GPCR superfamily; for convenience, the bulk of the alignment has been excised, as indicated by '.....' at the end of each paragraph. The centre of the alignment (highlighted) is poorly defined; the algorithm has not coped sensibly with the numerous gaps arising from insertions accumulated in other family members (not shown), and consequently leaves many residues and residue pairs stranded in a meaningless sea of gaps.

By contrast with the family of closely related adrenergic receptors (Figure 7.1), within the superfamily, sequences are highly divergent and are of disparate lengths. Consequently, large numbers of gaps are needed to bring equivalent residues into the correct register; where insertions have occurred in excised sequences relative to those shown, the alignment shows 'apparently' redundant gaps. From the paragraphs depicted, it is clear that the receptors have accumulated many insertions, especially in the centre of the alignment, producing a considerably extended output by comparison with Figure 7.1. Automatic alignment algorithms cannot easily resolve such fuzzy regions, and therefore tend to leave many single residues and residue pairs stranded, or **widowed,** from the well-aligned core, as illustrated by the highlighted region spanning the second and third paragraphs. Such over-zealous gap insertion is the hallmark of automatic methods, leading to ill-defined regions within alignments, with little or no biological meaning.

This result particularly spotlights the danger of using unsupervised, iterative automatic alignment procedures (the basis of the Pfam approach); with each iteration, results have the potential to include false-positive matches, which with successive iterations produce more and more corrupt alignments; eventually a point is reached where random sequences will make equally good, or better, matches than distantly related true-positive family members.

Let us now consider an alignment of the same superfamily, this time taken from a database of manually derived alignments (the seed alignments for the PRINTS database). For comparison, Figure 7.3 depicts the same region of the GPCR alignment shown in Figures 7.1 and 7.2 (again focusing on a portion that largely includes adrenergic receptors, together with receptors for dopamine and octopamine). By contrast with the automatically derived alignment illustrated in Figure 7.2, this hand-edited result endeavours to constrain gap insertion in such a way that residue widows are avoided and the integrity of the alignment core is not compromised. The resulting alignment is therefore more compact and, hopefully, conveys greater biological meaning.

Compare, for example, the well-conserved region highlighted in the second paragraph. The equivalent region is similarly highlighted in Figure 7.2, where it looks strikingly different, fragmented as it is by so many insertions. Thus, while it is evident that the alignment shown in Figure 7.3 is characterised by two highly conserved domains (which, in fact, are membrane-spanning, and probably α-helical), it is difficult to draw similar conclusions from Figure 7.2 – although there are two regions where gap insertion is less abundant, nevertheless no ungapped motif longer than seven residues can be found.

The differences seen in these alignments are extremely important. The contrasting results shown in Figures 7.1 and 7.2 demonstrate that automatic alignment of similar sequences is generally reliable, but that once distantly related sequences are included, the ability to produce faithful biological representations is compromised. This conclusion is supported by the striking differences observed between the alignments shown in Figures 7.2 and 7.3, where it is clear that the most reliable route to a biologically sensible result is via manual editing. This has important consequences for secondary database derivation, when we consider that the illustrated alignments are drawn from publicly available pattern databases, both of which purport to provide reliable

```
OAR_DROME    PINYAQ...KRTVGRVLLLISGVWLLSLLISSP.PLIG.WND.....WPDEFTSATP...
D2DR_RAT     PMLYNTR..YSSKRRVTVMIAIVWVLSFTISCP.LLFG.LNN...T.DQNE.........
D3DR_RAT     PVHYQHGTGQSSCRRVALMITAVWVLAFAVSCP.LLFG.FNT...TGDPSI.........
DADR_RAT     PFQYER...KMTPKAAFILISVAWTLSVLISFI.PVQLSWHKAK.PTWPLDGNFTSLEDT
DBDR_RAT     PFRYER...KMTQRVALVMVGLAWTLSILISFI.PVQLNWHRDKAGSQGQEGLLSNGTPW
A1AB_RAT     SLQYPT...LVTRRKAILALLSVWVLSTVISIG.PLLG.WKE......PAPNDDKE....
B1AR_RAT     PFRYQS...LLTRARARALVCTVWAISALVSFL.PILMHWW.....RAESD.EARRCYND
B2AR_HUMAN   PFKYQS...LLTKNKARVIILMVWIVSGLTSFL.PIQMHWY.....RATHQ.EAINCYAN
B2AR_RAT     PFKYQS...LLTKNKARVVILMVWIVSGLTSFL.PIQMHWY.....RATHK.QAIDCYAK
B3AR_RAT     PLRYGT...LVTKRRARAAVVLVWIVSATVSFA.PIMSQWW.....RVGADAEAQECHSN
5HTA_RAT     PIDYVN...KRTPRRAAALISLTWLIGFLISIP.PMLG.WRTPEDRSDPDA.........
5HTD_RAT     ALEYSK...RRTAGHAAAMIAAVWAISICISIP.PLF..WRQ..ATAHEEMSD.......
5HT2_RAT     PIHHSR...FNSRTKAFLKIIAVWTISVGISMPIPVFG.LQDDSKVFKEGS.........
....         .....


OAR_DROME    ...........CELTSQRG.................YVIYSSLGSFFIPLAIMTIVYIEIFVATRRRLRERARANK
D2DR_RAT     ...........CIIANPA.................FVVYSSIVSFYVPFIVTLLVYIKIYIVLRKRRKR......
D3DR_RAT     ...........CSISNPD.................FVIYSSVVSFYVPFGVTVLVYARIYIVLRQRQRK......
DADR_RAT     ED.......DNCDTRLSRT................YAISSSLISFYIPVAIMIVTYTSIYRIAQKQIRR......
DBDR_RAT     EEGWELEGRTENCDSSLNRT..............YAISSSLISFYIPVAIMIVTYTRIYRIAQVQIRR......
A1AB_RAT     ...........CGVTEEPF................CALFCSLGSFYIPLAVILVMYCRVYIVAKRTTKN......
B1AR_RAT     PK.........CCDFVTNRA...............YAIASSVVSFYVPLCIMAFVYLRVFREAQKQVKK......
B2AR_HUMAN   ET.........CCDFFTNQA...............YAIASSIVSFYVPLVIMVFVYSRVFQEAKRQLQK......
B2AR_RAT     ET.........CCDFFTNQA...............YAIASSIVSFYVPLVVMVFVYSRVFQVAKRQLQK......
B3AR_RAT     PR.........CCSFASNMP...............YALLSSSVSFYLPLLVMLFVYARVFVVAKRQRRF......
5HTA_RAT     ...........CTISKDHG................YTIYSTFGAFYIPLLLMLVLYGRIFRAARFRIRK......
5HTD_RAT     ...........CLVNTSQIS...............YTIYSTCGAFYIPSILLIILYGRIYVAARSRILN......
5HT2_RAT     ...........CLLADDN.................FVLIGSFVAFFIPLTIMVITYFLTIKSLQKEATL......
....         .....
```

Figure 7.3 Excerpt from an alignment of 687 G-protein-coupled receptors (GPCRs) taken from the PRINTS database. The result depicts the same part of the alignment as shown in Figures 7.1 and 7.2, but the alignment has been extended to include other members of the GPCR superfamily; for convenience, the bulk of the alignment has been excised, as indicated by '.....' at the end of each paragraph. Part of the figure, corresponding to the centre of the alignment shown in Figure 7.2, is highlighted: although this appears to be a poorly defined region in the automatically generated result, here it is evident that this region is in fact well conserved. Manual editing clearly allows sensible, restrained gap insertion, and helps to preserve the biological integrity of the alignment.

gene family discriminators. On the evidence of these alignments, the power of discrimination from the automatically generated alignment must be doubtful.

7.10 Searching databases with multiple alignments

In the previous chapter, we ended by considering different methods of searching a database for similarities to a single sequence query. Comparable approaches may be taken with multiple sequence alignments, which, as we have seen, may be processed into concise family descriptors or patterns (such as regular expressions, profiles, blocks, fingerprints or HMMs). Various computational techniques are used to search these complex data structures against a database, some of which are described in more detail in Chapter 8.

The advantage of using representations of multiple sequence alignment data in database searches is that more information is used, resulting in higher sensitivity (improved signal-to-noise ratio) compared with pairwise searches. The disadvantage, or trade-off, is that such searches take longer to run and the results are often more difficult to interpret.

In fact alignment searches tend to be performed only when pairwise searches have either returned already-known gene family members or provided no plausible leads at all. Alignment or pattern searching is then the next logical step to finding biologically significant matches in the Twilight Zone of sequence similarity.

7.10.1 PSI-BLAST

A recent, hybrid approach, incorporating elements of both pairwise and multiple sequence alignment methods, is Position-Specific Iterated or PSI-BLAST (Altschul *et al.*, 1997). While it is generally recognised that motif searches are the most sensitive and selective, able to detect weak but biologically meaningful similarities, their principal drawback is that methods to derive diagnostic family motifs can be very time-consuming and demand levels of understanding that render them inappropriate for general use. The innovation of the PSI-BLAST extension is that, following an initial database search, it allows automatic creation of position-specific profiles from groups of results that match the query above a defined threshold. Running the program several times can further refine the profile and increase search sensitivity.

As with other iterative methods, however, PSI-BLAST is not a complete solution, but has disadvantages as well as its many advantages. For example, unless low-complexity segments are masked, automated iterative searches have a tendency to degenerate to compositionally biased sequences (such as collagen, homopolymers, etc.), leading to profile dilution (Holm, 1998). Essentially, once a false sequence has crept into a profile, the search will thereafter be biased to accept many more unrelated sequences (see Section 7.9). It is therefore essential to inspect and validate homology relationships inferred from unsupervised iterative profile searches in order to be able to eliminate erroneous matches.

In Chapters 6 and 7, we have examined individual methods for pairwise and multiple sequence alignment, and have seen some of the pitfalls of blind reliance on purely automated approaches. In the following chapter, we review, in more detail, how multiple sequence information can be used to generate potent descriptors of family relationships, once again considering the relative pros and cons of both manual and automatic approaches.

7.11 Summary

- Analysis of groups of sequences that form gene families requires the ability to make connections between more than two family members. Multiple alignments are used to reveal conserved family characteristics.

- Multiple, like pairwise, alignments are simply models. There is nothing inherently correct or incorrect about a particular alignment. The important point is whether the model accurately reflects known biological data.

- Sequence- and structure-based alignments are both imperfect models, since neither can reflect all levels of biological information. Both approaches are valid representations of particular aspects of biology, and neither should therefore be considered to represent some ultimate truth or gold standard.

- A multiple alignment can be defined as a 2D table in which the rows represent individual sequences, and the columns the residue positions. A residue position within an unaligned sequence is termed the absolute position, while the aligned residue position is termed the relative position.

- The time taken to compute an alignment rises exponentially with the number of sequences to be aligned. Some methods use heuristics to reduce the time to find good (not necessarily optimal) alignments.

- Manual methods are often dismissed as being subjective. However, the results of automatic alignment programs almost invariably require manual polishing, and hence alignment editors have become essential tools.

- Simultaneous multiple alignment methods align all sequences within a set at once, and hence are very time-consuming; they work best on small sets of short sequences.

- Progressive multiple alignment methods align sequences in pairs, following the branching order of a family tree. The most similar are aligned first, and more distantly related sequences are added later. By exploiting likely evolutionary relationships, such methods can handle more realistic data-sets in a timely and cost-effective manner.

- There are numerous alignment databases accessible via the Web. These result from different approaches: e.g., the application of automated methods to cluster the primary sequence resources into families, or from endeavours to produce gene family discriminators for inclusion in secondary databases.

- Alignments produced by purely automatic (especially iterative) methods should be handled with care, especially in cases where sequence similarity is low; they often result in over-zealous gap insertion and can produce misalignments.

- Various computational techniques have evolved to search primary sequence databases using alignment-based data structures. A recent hybrid approach, incorporating elements of both pairwise and multiple alignment methods, is Position-Specific Iterated or PSI-BLAST. Although fast to run, it has the disadvantage that the automated iterative search may degenerate and lead to profile dilution.

7.12 Further reading

Database search methods
ALTSCHUL, S.F., MADDEN, T.L., SCHÄFFER, A.A., ZHANG, J., ZHANG, Z., MILLER, W. and LIPMAN, D.J. (1997) Gapped BLAST and PSI-BLAST: a new generation of protein database search programs. *Nucleic Acids Research*, **25**(17), 2289–3402.

HOLM, L. (1998) Unification of protein families. *Current Opinion in Structural Biology*, **8**, 372–379.

Alignment methods
FENG, D.-F. and DOOLITTLE, R.F. (1987) Progressive sequence alignment as a prerequisite to correct phylogenetic trees. *Journal of Molecular Evolution*, **25**, 351–360.

FENG, D.-F. and DOOLITTLE, R.F. (1996) Progressive alignment of amino acid sequences and construction of phylogenetic trees from them. *Methods in Enzymology*, **266**, 368–382.

THOMPSON, J.D., HIGGINS, D.G. and GIBSON, T.J. (1994) CLUSTAL W: Improving the sensitivity of progressive multiple sequence alignment through sequence weighting, position-specific gap penalties and weight matrix choice. *Nucleic Acids Research*, **22**(22), 4673–4680.

CHAPTER EIGHT

Secondary database searching

8.1 Introduction

Building on the themes of primary database searching and multiple sequence alignment, this chapter deals with the analysis methods that underlie secondary database searches. We introduced secondary databases in a general way in Chapter 3, highlighting aspects of database content and format. Here, we describe in more detail the types of information stored in the major resources, including regular expressions, profiles, fingerprints, blocks and Hidden Markov Models. The aim is to provide an understanding of the principal diagnostic strengths and weaknesses of the different techniques, which must always be borne in mind when interpreting search results.

8.2 Why bother with secondary database searches?

Before further exploring secondary database diagnostic methods, in light of the significant advances in primary database search technology (such as BLAST, and especially its iterative extension, PSI-BLAST), it is worth reflecting for a moment why secondary searches might still be useful.

As we have seen, primary databases are growing at an alarming rate; and much of current research is focused on the ability to deduce functional features in the wealth of their as yet uncharacterised sequences by recognising relationships with known ones. Primary database search tools are effective for identifying sequence similarities, but analysis of output is sometimes difficult and cannot always answer some of the more sophisticated questions of sequence analysis. There are many reasons for this: e.g., in 1998, GenBank contained more than a million sequences from more than 18 000 organisms, resulting in complex and redundant search outputs; unless masking devices are used, results can be dominated by irrelevant matches to low-complexity sequences; the presence of highly repetitive, modular sequences can also greatly complicate interpretation; similarly, with multi-domain proteins, it may not be clear at what level a match has been made (e.g., at the level of a single domain, of several domains, or of the whole protein); truncated description lines of matched database sequences may be uninformative or, at worst, ambiguous, leading to possible errors when making functional inferences; and, given database size and increasing levels of noise, similarity between orthologous sequences may not be as high as that between sequences that do not belong to the same gene family, and related sequences may hence fail to attain significant scores.

Orthology provides an important layer of information when considering phylogenetic relationships between genes. We are thus beginning to see a change of emphasis in sequence analysis, from attempts simply to infer homology, to the more exacting task of recognising orthology (Huynen and Bork, 1998). Secondary databases may offer a small step in this direction. Depending on the type of analysis method used, relationships may be elucidated in considerable detail, including superfamily, family, subfamily, and species-specific sequence levels. The ability to distil sequence information in such precise ways makes secondary database searching a useful and often powerful adjunct to routine primary searches.

8.3 What's in a secondary database?

In Chapter 3, we introduced a number of secondary databases and mentioned the different types of information they contain. It is important not only to understand the different data types, but also to know how to search them, how to interpret the range of different outputs, and how to assess the biological significance of the results.

You will recall that the underlying principle behind the development of secondary databases is that within multiple alignments can be found conserved motifs that reflect shared structural or functional characteristics of the constituent sequences – see Figure 8.1. Such conserved motifs, or indeed the complete parent alignments, may be used to build characteristic signatures that aid family and/or functional diagnoses of newly determined sequences.

Different secondary databases have evolved as a result of the different analysis methods used in the derivation of family signatures. Some of the main approaches are outlined below.

8.3.1 Regular expressions

The simplest approach to pattern recognition is to characterise a family by means of a single conserved motif, and to reduce the sequence data within the motif to a consensus or regular expression pattern. Regular expressions, then, discard sequence data, retaining only the most conserved or significant residue information, as shown in Table 8.1.

The expression derived from the motif shown in Table 8.1 indicates that positions 2, 4, 10, 12 and 15 are completely conserved; positions 1, 11, 13 and 14 allow one of two possible residues (i.e., residues specified within square brackets are allowed at that position); position 3 allows one of three

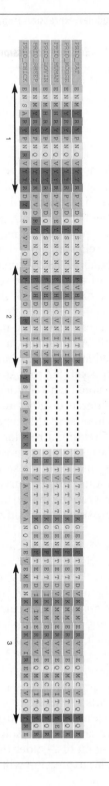

Figure 8.1 Colour-coded sequence alignment – gap insertion, bringing equivalent parts of the alignment into the correct register, leads to the formation of islands of conservation, or motifs (denoted by the arrows). Motifs may be encoded in different ways to generate family-specific discriminators (e.g., regular expressions, fingerprints, profiles, etc.).

Table 8.1 Derivation of a regular expression from a conserved motif.

Alignment	Regular expression
ADLGAVFALCDRYFQ	
SDVGPRSCFCERFYQ	[AS]-D-[IVL]-G-x4-{PG}-C-[DE]-R-[FY]2-Q
ADLGRTQNRCDRYYQ	
ADIGQPHSLCERYFQ	

possible residues; positions 5 to 8 can be anything (as denoted by x4); and position 9 can be anything except proline or glycine (i.e., residues specified within curly brackets are disallowed at that position).

In order to reduce the likelihood of a pattern making too many incorrect matches, the software that makes use of regular expressions often does not tolerate similarity, and searches are thus limited in scope to the retrieval of *identical* matches. For example, let us suppose that a query sequence matches the expression in Table 8.1 at all but the second position, where it has a glutamic acid conservatively substituting for the prescribed aspartic acid. Such a sequence, in spite of being 99% identical to the expression, will nevertheless be rejected as a mismatch, even though the mismatch is a conservative, biologically feasible replacement. Alternatively, a sequence matching all positions of the pattern, but with an additional residue inserted in the non-conserved region following the glycine, will again fail to match, because the expression does not cater for sequences with more than 4 'linking' residues at this point. Searching a database in this way thus results in either an exact match, or no match at all.

The strict binary outcome of this type of pattern searching has severe diagnostic limitations. Creating a regular expression that performs well in database searches is always a compromise between the tolerance that can be built into it, and the amount of noise it will match: the fuzzier the pattern, the noisier its results, but the greater the hope of finding distant relatives; while the stricter the pattern, the cleaner its results, but the greater the chance of missing true-positive matches not catered for within the defined expression.

A further limitation of this approach hinges on the philosophy of using single motifs to characterise entire protein families. For best results, this effectively requires us to know in advance what is the most conserved region of a sequence alignment. In some cases, as with enzyme active sites, for example, the choice may be clear-cut. However, for families where this kind of information is not available, there may be a number of conserved regions, only one of which, let us say, is chosen to create a regular expression. The problem is that the parent alignment in which the conserved motifs are identified is a reflection only of the sequences in the current primary sources. As the source databases grow, and new sequences become available, these alignments are likely to evolve. As a result, in some cases, the region of the alignment originally used to characterise the particular protein family may have changed considerably, and is ultimately less conserved than neighbouring motifs. Under these circumstances, the diagnostic performance of the regular expression will fall off with time, and the pattern must be modified (or changed completely) to more accurately reflect the contents of the primary databases.

Rules

Regular expressions are used to best effect when a particular protein family can be characterised by a highly conserved motif (typically 10–20 residues in length), which may perhaps be diagnostic of some core piece of the protein architecture or of some critical functional role. Often, however, it is possible to identify much shorter, generic patterns within sequence alignments that are not associated with specific protein families.

Sequence features of this sort are believed to be the result of convergence to a common property: e.g., they may denote sugar attachment sites, phosphorylation or hydroxylation sites, and so on. Such patterns, which may be as short as 3–4 residues in length (see Table 8.2), cannot be used for family diagnosis, and do not provide good discrimination: this is because the shorter the motif, the greater the chance of random matches to it. For example, in OWL29.6 there were **71** exact matches to the sequence motif Asp-Ala-Val-Ile-Asp (DAVID), and **1088** exact matches to the shorter form Asp-Ala-Val-Glu (DAVE). Thus, short motifs are diagnostically unreliable (because they are non-specific), and matches to them, in isolation, are relatively meaningless. Realistically, short motifs can only be used to provide a guide as to whether a certain type of functional site *might* exist in a sequence, which information must be verified by experiment. Such patterns are termed **rules** to distinguish them from family-specific regular expressions.

Table 8.2 Example functional sites with the regular expression rules used to detect them.

Functional site	Rule
N-glycosylation site	N-{P}-[ST]-{P}
Protein kinase C phosphorylation site	[ST]-x-[RK]
Casein kinase II phosphorylation site	[ST]-x(2)-[DE]
Asp and Asn hydroxylation site	C-x-[DN]-x(4)-[FY]-x-C-x-C

Regular expressions (patterns and rules) are the basis of the PROSITE database. The diagnostic problems outlined above have recently led to the inclusion into this database of alternative discriminators where patterns are most likely to fail. These are profiles, which are discussed in more detail later.

Fuzzy regular expressions

One response to the strict nature of regular-expression pattern matching is to build an element of tolerance or fuzziness into the patterns. As we saw in Chapter 3, one approach is to consider amino acid residues as members of groups, as defined by shared biochemical properties: e.g., FYW are aromatic, HKR are basic, ILVM are hydrophobic, and so on (see Table 3.4 and Figure 8.2).

Taking such biochemical properties into account, the motif depicted in Table 8.1 yields the more permissive regular expression shown in Table 8.3.

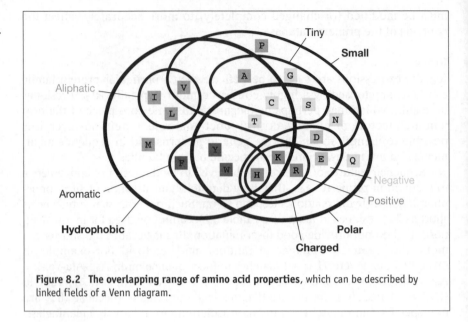

Figure 8.2 The overlapping range of amino acid properties, which can be described by linked fields of a Venn diagram.

Table 8.3 Fuzzy regular expression representation of a conserved motif.

Alignment	Fuzzy regular expression
ADLGAVFALCDRYFQ	
SDVGPRSCFCERFYQ	[ASGPT]-D-[IVLM]-G-X5-C-[DENQ]-R-[FYW]2-Q
ADLGRTQNRCDRYYQ	
ADIGQPHSLCERYFQ	

Expressions of this type are exploited by the eMOTIF system, which takes as its data sources alignments in the BLOCKS and PRINTS databases. It should be clear from this simple example that such patterns are more relaxed, accepting a wider range of residues at particular positions. This has the potential advantage of being able to recognise more distant relatives, but has the inherent disadvantage that it will also match many more sequences simply by chance. For example, let us return for a moment to the motif DAVID, which we saw had 71 exact matches in OWL29.6 (see Table 8.4). If we introduce one fuzzy position (e.g., let us say that the final D may belong to the group DEQN), then we find 252 matches in this version of the database; with two fuzzy positions, we retrieve 925 matches; with three, the number increases to 2739; and with tolerance at all five positions, **51 506** matches are retrieved! Clearly, the more tolerant each position within a motif with regard to the types of residue allowed, the more permissive the resulting regular expression; and the shorter the motif (as with PROSITE rules), the worse the situation becomes.

Because a pattern effectively represents the minimum expression of an aligned motif, sequence information is lost and parts of it become ill-defined. The more divergent the aligned sequences, the fuzzier the pattern becomes,

Table 8.4 Illustration of the effects of introducing fuzziness into regular expressions, and of motif length, on the number of matches retrieved from sequence database searches.

151

What's in a secondary database?

Regular expression	No. of exact matches (OWL29.6)
D-A-V-I-D	71
D-A-V-I-[DENQ]	252
[DENQ]-A-V-I-[DENQ]	925
[DENQ]-A-[VLI]-I-[DENQ]	2739
[DENQ]-[AG]-[VLI]2-[DENQ]	51506
D-A-V-E	1088

and the more likely the expression is to make false-positive matches. Results of searching with regular expressions must therefore be interpreted with care – if a query sequence matches an expression, there is no guarantee that the match is biologically meaningful; conversely, if a sequence *fails* to match an expression, it does not necessarily mean that the query is not a family member (as we have seen, it may simply be that it deviates from the pattern by a single residue). In other words, matches to regular expressions are not necessarily true, and mismatches are not necessarily false. To address some of the diagnostic limitations of patterns, more sophisticated approaches have been devised to improve diagnostic performance and separate out biologically meaningful matches from the sea of noise that large databases effectively represent.

8.3.2 Fingerprints

Within a sequence alignment, it is usual to find not one, but several motifs that characterise the aligned family. Diagnostically, it makes sense to use many, or all, of the conserved regions to create a signature or fingerprint, so that, in a database search, there is a higher chance of identifying a distant relative, whether or not all parts of the signature are matched. In one such approach, groups of motifs are excised from alignments, and the sequence information they contain is converted into matrices populated only by the residue frequencies observed at each position of the motifs, as illustrated in Table 8.5. This type of scoring system is said to be *unweighted,* in the sense that no additional scores (e.g., from mutation or substitution matrices) are used to enhance diagnostic performance.

In the example shown in Table 8.5(a), the motif is 13 residues long and 12 sequences deep. The maximum score in the resulting frequency matrix (Table 8.5(b)) is thus 12. Rapid inspection of the matrix indicates that positions 7, 9 and 12 are completely conserved, corresponding to Lys, Leu and Pro residues respectively. Any residue not observed in the motif takes no score – the matrix is thus sparse, with few positions scoring and most with zero score.

The use of raw residue frequencies has not been a popular approach because their scoring potential is relatively limited and, for motifs containing only few sequences, the ability to detect distant homologues is compromised because the matrices do not contain sufficient variation.

Table 8.5 Example illustrating (a) an ungapped, aligned motif; and (b) its corresponding frequency matrix, based on residues observed at each position in the motif (each motif column corresponds to a row in the matrix).

(a)

```
YVTVQHKKLRTPL
YVTVQHKKLRTPL
YVTVQHKKLRTPL
AATMKFKKLRHPL
AATMKFKKLRHPL
YIFATTKSLRTPA
VATLRYKKLRQPL
YIFGGTKSLRTPA
WVFSAAKSLRTPS
WIFSTSKSLRTPS
YLFSKTKSLQTPA
YLFTKTKSLQTPA
```

(b)

T	C	A	G	N	S	P	F	L	Y	H	Q	V	K	D	E	I	W	R	M	B	X	Z
0	0	2	0	0	0	0	0	0	7	0	0	1	0	0	0	0	2	0	0	0	0	0
0	0	3	0	0	0	0	0	2	0	0	0	4	0	0	0	3	0	0	0	0	0	0
6	0	0	0	0	0	0	6	0	0	0	0	0	0	0	0	0	0	0	0	0	0	0
1	0	1	1	0	3	0	0	1	0	0	0	3	0	0	0	0	0	0	2	0	0	0
2	0	1	1	0	0	0	0	0	0	0	3	0	4	0	0	0	0	1	0	0	0	0
4	0	1	0	0	1	0	2	0	1	3	0	0	0	0	0	0	0	0	0	0	0	0
0	0	0	0	0	0	0	0	0	0	0	0	0	12	0	0	0	0	0	0	0	0	0
0	0	0	0	0	6	0	0	0	0	0	0	0	6	0	0	0	0	0	0	0	0	0
0	0	0	0	0	0	0	0	12	0	0	0	0	0	0	0	0	0	0	0	0	0	0
0	0	0	0	0	0	0	0	0	0	0	2	0	0	0	0	0	0	10	0	0	0	0
9	0	0	0	0	0	0	0	0	0	2	1	0	0	0	0	0	0	0	0	0	0	0
0	0	0	0	0	0	12	0	0	0	0	0	0	0	0	0	0	0	0	0	0	0	0
0	0	4	0	0	2	0	0	6	0	0	0	0	0	0	0	0	0	0	0	0	0	0

Nevertheless, frequency matrices have been used to good effect as part of the fingerprinting technique used as the basis for the PRINTS database.

As discussed in Chapter 3, in creating a fingerprint, discriminating power is enhanced by iterative database scanning. The motifs therefore grow and become more mature with each database pass, as more sequences are matched and further residue information is included in the matrices. In real terms, this successively shifts the seed frequencies towards more representative values as the fingerprint incorporates additional family members. For example, after four iterations, the motif shown in Table 8.5(a) has grown to a depth of 73 sequences; as seen in Table 8.6(a), at the end of this process, only position 9 remains conserved, and more variation is observed in the matrix as a whole. Nevertheless, although clearly more densely populated than the initial frequency matrix, this matrix is still relatively sparse – there are still more

Table 8.6 (a) The frequency matrix derived from the initial motif shown in Table 8.5(a) after four database iterations; and (b) the PAM-weighted matrix derived from the same motif.

(a)

T	C	A	G	N	S	P	F	L	Y	H	Q	V	K	D	E	I	W	R	M	B	X	Z
0	0	4	0	0	0	0	8	4	34	0	0	15	0	0	0	1	7	0	0	0	0	0
0	4	15	0	0	0	0	0	7	0	0	0	37	0	0	0	10	0	0	0	0	0	0
50	0	0	0	0	3	0	18	0	0	0	0	0	0	0	0	0	0	0	2	0	0	0
3	0	12	2	1	8	0	3	6	0	0	0	14	0	0	0	15	2	0	7	0	0	0
9	2	2	2	1	1	0	0	0	0	1	25	0	20	0	6	0	0	4	0	0	0	0
14	0	2	0	0	4	0	14	0	8	31	0	0	0	0	0	0	0	0	0	0	0	0
0	0	1	0	0	0	0	0	0	0	0	0	0	70	0	0	0	0	2	0	0	0	0
0	0	2	1	0	17	0	0	0	0	0	0	0	52	0	0	0	0	1	0	0	0	0
0	0	0	0	0	0	0	0	73	0	0	0	0	0	0	0	0	0	0	0	0	0	0
0	0	0	0	0	0	0	0	0	0	0	0	5	0	0	0	0	0	68	0	0	0	0
44	0	0	0	0	6	0	0	0	0	12	11	0	0	0	0	0	0	0	0	0	0	0
0	0	1	0	0	0	69	0	0	0	3	0	0	0	0	0	0	0	0	0	0	0	0
2	0	11	0	0	7	0	0	53	0	0	0	0	0	0	0	0	0	0	0	0	0	0

(b)

T	C	A	G	N	S	P	F	L	Y	H	Q	V	K	D	E	I	W	R	M
-29	-22	-29	-48	-24	-24	-46	40	-13	62	-10	-40	-22	-38	-44	-44	-15	16	-30	-22
-1	-32	-1	-18	-20	-10	-13	-9	20	-22	-21	-18	32	-23	-22	-20	32	-61	-26	19
0	-36	-18	-30	-24	-12	-30	36	0	24	-18	-36	-6	-30	-36	-30	6	-30	-30	-6
3	-29	3	-4	-10	-1	-7	-22	3	-31	-19	-15	14	-12	-15	-13	11	-52	-15	11
3	-48	-1	-8	7	1	-4	-54	-31	-46	6	14	-17	23	6	5	-20	-48	14	-9
2	-27	-7	-19	-3	-5	-13	0	-16	6	8	-10	-11	-15	-13	-11	-7	-37	-12	-15
0	-60	-12	-24	12	0	-12	-60	-36	-48	0	12	-24	60	0	0	-24	-36	36	0
6	-30	0	-6	12	12	0	-48	-36	-42	-6	0	-18	30	0	0	-18	-30	18	-12
-24	-72	-24	-48	-36	-36	-36	24	72	-12	-24	-24	24	-36	-48	-36	24	-24	-36	48
-12	-50	-20	-32	2	-2	0	-50	-34	-48	26	18	-24	32	-6	-6	-24	10	62	-2
24	-29	7	-5	5	6	0	-36	-24	-31	6	1	-6	1	4	4	-6	-56	-4	-14
0	-36	12	-12	-12	12	72	-60	-36	-60	0	0	-12	-12	-12	-12	-24	-72	0	-24
-6	-44	-2	-18	-16	-10	-12	-10	22	-24	-18	-14	10	-22	-24	-18	6	-40	-26	16

zero than there are scoring positions. In database searches, therefore, this matrix will perform cleanly (with little noise) and with high specificity.

This is to be contrasted with a situation where, for example, a PAM matrix is used to weight the scores, in order to allow more distant relationships to be recognised. The effect of weighting the initial matrix in this way is illustrated in Table 8.6(b). As can be seen, the PAM-weighted result, even though based on the initial sparse matrix, is highly populated. Consequently, in database searches, experience shows that such a matrix will perform with high levels of noise and relatively low specificity. A distant relative might well achieve a higher score with this approach, but so inevitably will random matches, as residues that are not observed in the initial (or indeed in the *final*) motifs are given significant weights. Compare, for example, the ninth motif position, which even after four database iterations had remained conserved; in the PAM-weighted matrix, this position is no longer completely conserved, four other residues (Phe, Val, Ile and Met) being assigned large positive scores.

Because of the poor signal-to-noise performance of weighted matrices, the technique of fingerprinting has adhered to the use of residue frequencies alone. Diagnostic performance is enhanced through the iterative process, but the full potency of the method is gained from the mutual context provided by motif neighbours. This is important, as the method inherently implies a biological context to motifs that are matched in the correct order in a query sequence, with appropriate intervals between them (see Box 8.1). This allows sequence identification even when some parts of a fingerprint are absent. Thus, for example, a sequence that matches only four of seven motifs may still be diagnosed as a true match if the motifs are matched in the correct order and the distances between them are consistent with those expected of true neighbouring motifs. This situation is illustrated in Figure 8.3, in which graphs are plotted for (a) a complete match, and (b) a partial, but nevertheless true, match to the G-protein-coupled receptor fingerprint.

8.3.3 Blocks

As mentioned above, the constituent motifs of a fingerprint are unweighted, which sometimes compromises diagnostic performance. Nevertheless, bearing in mind signal-to-noise issues, it is possible to build alternative motif representations by applying different weighting schemes.

One such approach is embodied in the BLOCKS database. Here, conserved motifs, or blocks, are located by searching for spaced residue triplets (e.g., Ala-x-x-x-Val-x-x-Cys, where x represents any amino acid), and a block score is calculated using the BLOSUM 62 substitution matrix. The validity of blocks found by this method is confirmed by the application of a second motif-finding algorithm, which searches for the highest-scoring set of blocks that occur in the correct order without overlapping. Blocks found by both methods are considered to be more reliable and are entered in the database.

A typical block is shown in Figure 8.4. Sequence segments that comprise the block are clustered to reduce multiple contributions to residue frequencies from groups of closely related sequences. This is particularly important for very deep motifs (i.e., with contributions from tens or hundreds of sequences), which can be dominated by numerous virtually identical sequences, reflecting the innate bias in the primary databases. With this approach, each cluster is treated as a single segment, each of which is assigned a score that gives a measure of its relatedness. The higher the weight, the more dissimilar the segment is from other segments in the block; the most distant segment is given a weight of 100. Sequence fragments that are less than 80% similar are separated by blank lines, as shown in Figure 8.4.

In a manner analogous to protein fingerprinting, blocks may be used to search sequence databases to find additional family members. Blocks within a family are converted to position-specific substitution matrices (PSSMs – pronounced possums), which are used to make independent database searches. The results are compared and, where more than one block detects the same

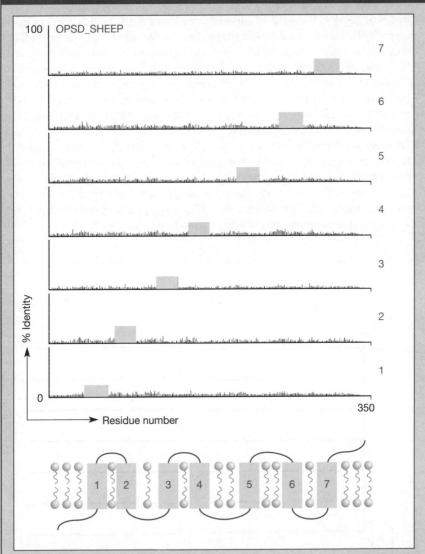

Protein fingerprints are groups of motifs that represent the most conserved regions of multiple sequence alignments. Features of alignments are conserved usually because they are critical to the function or structure of the protein: e.g., they may contain the catalytic residues within an enzyme active site; or they may encode strands or helices vital to the protein scaffold, providing, in a sense, the architectural blueprints for particular protein folds. Consider the structure of rhodopsin, a member of the 7TM or serpentine receptors, so-called because they are believed to comprise seven membrane-spanning helices that snake back-and-forth across the membrane. Sequence alignments show the TM regions to be sufficiently well conserved to provide a characteristic signature for rhodopsin-like receptors. It is therefore possible to construct a

BOX 8.1: CONTINUED

fingerprint based on the seven TM domains. Visualising a match between a known fin-gerprint and an unknown query sequence involves the creation of a graph in which the query sequence is plotted along the *x*-axis, and scores with the constituent motifs of the fingerprint are plotted on the *y*-axis. For each fingerprint element, a window corresponding to the width of the motif is slid across the query sequence, and an identity score calculated using the associated frequency matrix derived from the motifs stored in the PRINTS database. Where a window scores above a given threshold, a block is used to mark the position of the match. For a true-positive match to a rhodopsin-like 7TM receptor, we would expect to see seven consecutive blocks, from the N- to the C-terminus, corresponding to each of the seven TM domains. Fingerprint graphs of this type facilitate rapid visual inspection of queries, allowing ready identification of complete matches, with all seven TM domains, and of partial matches and fragments that lack matches with one or more motifs. The ability to recognise both complete and partial matches renders protein fingerprinting a powerful diagnostic tool.

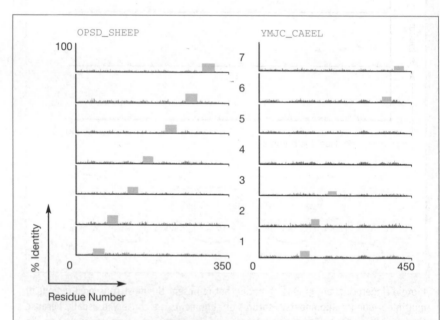

Figure 8.3 Graphs used to visualise protein fingerprints. The horizontal axis represents the query sequence, the vertical axis the % score of each motif (0–100 per motif), and each block a residue-by-residue match in the sequence, its leading edge marking the first position of the match. The profiles depict rhodopsin-like GPCR fingerprints of ovine rhodopsin and of a *C. elegans* hypothetical protein. Solid blocks appearing in a systematic order along the sequence and above the level of noise indicate matches with the constituent motifs. Ovine rhodopsin is a known true-positive family member, matching all seven motifs; the *C. elegans* sequence fails to make a complete match, but can still be identified with the GPCR superfamily because of the diagnostic framework provided by the five well-matched motif neighbours.

```
CCKR_HUMAN ( 362) SSCVNPIIYCFMNKRFR   3
 CCKR_RAT ( 378) SSCVNPIIYCFMNKRFR   3

FML2_HUMAN ( 294) NSCLNPMLYVFVGQDFR   4
FMLR_HUMAN ( 293) NSCLNPMLYVFMGQDFR   4
FMLR_MOUSE ( 304) NSCLNPMLYVFMGQDFR   4
FMLR_RABIT ( 295) NSCLNPMLYVFMGQDFR   4

GASR_CANFA ( 388) SACVNPLVYCFMHRRFR   5
GASR_HUMAN ( 382) SACVNPLVYCFMHRRFR   5
GASR_PRANA ( 385) SACVNPLVYCFMHRRFR   5
GASR_RABIT ( 387) SACVNPLVYCFMHRRFR   5
 GASR_RAT ( 387) SACVNPLVYCFMHRRFR   5

ET1R_BOVIN ( 361) NSCINPIALYFVSKKFK   9
 ET1R_RAT ( 361) NSCINPIALYFVSKKFK   9
ETBR_BOVIN ( 377) NSCINPIALYLVSKRFK   9
ETBR_HUMAN ( 378) NSCINPIALYLVSKRFK   9
 ETBR_PIG ( 379) NSCINPIALYLVSKRFK   9
 ETBR_RAT ( 378) NSCINPIALYLVSKRFK   9

OPSD_LOLFO ( 307) SAIHNPMIYSVSHPKFR  12
OPSD_OCTDO ( 308) SAIHNPIVYSVSHPKFR  12
OPSD_TODPA ( 306) SAIHNPMIYSVSHPKFR  12

P2UR_HUMAN ( 296) NSCLDPVLYFLAGQRLV  13
P2UR_MOUSE ( 298) NSCLDPVLYFLAGQRLV  13
 P2UR_RAT ( 297) NSCLDPVLYFLAGQRLV  13

  5H6_RAT ( 312) NSTMNPIIYPLFMRDFK  16

EDG1_HUMAN ( 302) NSGTNPIIYTLTNKEMR  21

EBI2_HUMAN ( 300) NCCMDPFIYFFACKGYK  23

OXYR_HUMAN ( 321) NSCCNPWIYMLFTGHLF  24
 OXYR_PIG ( 323) NSCCNPWIYMLFTGHLF  24
V1AR_HUMAN ( 340) NSCCNPWIYMFFSGHLL  18
 V1AR_RAT ( 346) NSCCNPWIYMFFSGHLL  18

PER3_BOVIN ( 337) NQILDPWVYLLLRKILL  35
PER3_HUMAN ( 338) NQILDPWVYLLLRKILL  35

YN84_CAEEL ( 331) SCVAYPLIFTLLNRGIR 100
```

Figure 8.4 Part of a block, in which sequence segments are clustered and weighted according to their relatedness – the most distant sequence within the block scores 100.

sequence, and the distances between the blocks are consistent with known family members, a probability (P) value is calculated for the multiple hit.

For a given sequence, as with a fingerprint, the more blocks matched, the greater the confidence that the sequence belongs to that family, provided the blocks are matched in the correct order and have appropriate distances between them. However, as with other weighting schemes, there is a diagnostic trade-off between the ability to capture all true-positive matches with a set of blocks, and the likelihood of making false-positive matches. A further complication arises with the assessment of significance of high-scoring individual blocks versus lower-scoring multiple-block matches. With the blocks approach, it is possible for high-scoring, biologically meaningless hits to score above lower-scoring, but nevertheless biologically significant, true-positive matches. Care is therefore required in the interpretation of BLOCKS search output, as the top hit is not necessarily always the correct one.

8.3.4 Profiles

By contrast with motif-based pattern recognition techniques, an alternative approach is to distil the sequence information within *complete* alignments into scoring tables, or profiles. Profiles define which residues are allowed at given positions; which positions are highly conserved and which degenerate; and which positions, or regions, can tolerate insertions. The scoring system is intricate, and may include evolutionary weights and results from structural studies, as well as data implicit in the alignment. In addition, variable penalties may be specified to weight against INDELs occurring within core secondary structure elements.

An example profile is shown in Figure 8.5. The I and M fields contain position-specific scores for insert and match positions respectively. They take the form:

```
/I: [ SY=char1; parameters; ]
/M: [ SY=char2; parameters; ]
```

where:

- char1 is a character representing an insert position in the parent alignment;

- char2 is a character representing a match position in the parent alignment; and

- parameters is a list of specifications assigning values to various position-specific scores (details of these parameters are outside the scope of this book, but they include initiation and termination scores, state transition scores, insertion/match/deletion extension scores, and so on).

In this example, we can see that the profile contains three conserved blocks separated by two gapped regions. Within the conserved blocks, although small insertions and deletions are not totally forbidden, they are strongly impeded by large gap penalties defined in the DEFAULT data block: MI = –26, I = –3, MD = –26, D = –3 (MI is a match-insert transition score, I is an insert extension score, MD is a match-delete transition score and D is a deletion extension score). These penalties are superseded by more permissive values in the two gapped regions (e.g., in the first, MI = 0, I = –1, MD = 0, etc.).

The inherent complexity of profiles renders them highly potent discriminators. They are therefore used to complement some of the poorer regular expressions in PROSITE, and/or to provide a diagnostic alternative where extreme sequence divergence renders the use of patterns inappropriate.

```
/DEFAULT: MI=-26; I=-3; IM=0; MD=-26; D=-3; DM=0;
/M: SY='F';M=-2,-3,-3,-4,2,-3,-2,1,-2,0,-1,-2,-3,-3,-4,-2,-1,0,-5,2;
/M: SY='I';M=-1,-5,-2,-3,-2,-3,0,1,1,-1,1,-1,-2,-1,-1,-1,0,1,-4,-4;
/M: SY='A';M=2,-3,1,0,-5,2,-2,-1,-1,-3,-2,1,1,0,-2,2,2,0,-8,-5;
/M: SY='L';M=-3,-8,-5,-4,2,-6,-2,2,-4,6,4,-3,-3,-2,-3,-3,-2,1,-3,0;
/M: SY='Y';M=-4,-2,-6,-6,9,-7,0,-1,-5,-1,-3,-3,-6,-5,-6,-4,-4,-4,-1,11;
/M: SY='D';M=1,-6,3,3,-7,0,0,-2,-1,-4,-3,2,0,1,-2,0,0,-2,-9,-6;
/M: SY='Y';M=-5,-3,-6,-6,10,-7,-1,-1,-2,-1,-2,-3,-6,-5,-5,-4,-4,-4,-1,11;
/M: SY='K';M=-1,-6,1,1,-4,-2,0,-2,2,-3,-1,1,-1,1,1,0,0,-3,-7,-6;
/M: SY='A';M=1,-4,1,0,-5,1,-1,-1,0,-3,-1,1,0,0,0,1,1,-1,-7,-6;
/M: SY='R';M=0,-5,0,0,-5,-1,0,-1,1,-3,-1,1,0,1,1,0,0,-2,-5,-5;
/M: SY='R';M=0,-5,1,1,-6,0,1,-2,1,-4,-2,1,0,1,2,1,0,-2,-5,-5;
/M: SY='E';M=1,-6,2,2,-6,0,0,-2,-1,-4,-2,1,1,1,-1,0,0,-3,-8,-6;
/M: SY='D';M=0,-6,2,2,-6,0,1,-3,0,-5,-3,2,-1,2,-1,0,0,-4,-7,-4;
/M: SY='D';M=0,-8,4,3,-6,0,0,-2,-1,-3,-2,2,-2,2,-2,0,-1,-3,-9,-6;
/M: SY='L';M=-2,-8,-5,-5,2,-5,-3,3,-4,7,5,-4,-3,-3,-4,-3,-2,3,-4,-2;
/M: SY='S';M=1,-4,1,1,-5,1,0,-2,1,-4,-2,1,0,0,0,1,1,-2,-6,-5;
/M: SY='F';M=-3,-7,-6,-6,6,-5,-3,3,-2,5,3,-4,-5,-4,-5,-4,-3,1,-3,3;
/M: SY='Q';M=-1,-6,0,0,-3,-2,1,-1,1,-2,0,0,-1,1,1,-1,0,-1,-6,-4;
/M: SY='K';M=-1,-8,0,1,-3,-2,0,-2,3,-3,0,1,0,2,2,0,0,-3,-6,-6;
/M: SY='G';M=2,-5,1,0,-7,7,-3,-4,-2,-6,-4,1,-1,-2,-4,2,0,-2,-10,-8;
/M: SY='D';M=1,-7,5,4,-8,1,1,-3,0,-5,-3,2,1,2,-2,0,0,-4,-10,-6;
/M: SY='I';M=0,-5,-1,-2,-2,-2,-1,2,0,0,1,-1,-2,0,0,-1,0,1,-6,-5;
/M: SY='L';M=-2,-6,-5,-5,3,-5,-3,4,-3,6,4,-4,-4,-3,-4,-3,-2,3,-5,0;
/M: SY='Q';M=-1,-5,-1,-1,-3,-2,0,0,0,-2,-1,0,-1,0,0,-1,0,-1,-6,-3;
/M: SY='V';M=0,-4,-3,-4,-1,-3,-3,5,-3,3,3,-2,-2,-2,-3,-2,0,5,-8,-4;
/M: SY='L';M=-1,-6,-3,-3,-1,-3,-2,2,-3,3,2,-2,-2,-2,-3,-2,-1,2,-5,-3;
/M: SY='D';M=0,-6,3,3,-6,0,1,-3,2,-5,-2,2,-1,2,1,0,0,-4,-7,-5;
/M: SY='K';M=-1,-6,0,0,-2,-1,0,-3,3,-4,-1,1,0,1,0,0,0,-3,-6,-4;
/M: SY='N';M=1,-4,1,1,-5,0,0,-2,0,-3,-2,1,1,0,-1,1,1,-1,-7,-5;
   /I: MI=0; I=-1; MD=0; /M: SY='X'; M=0; D=-1;
/M: SY='G';M=1,-5,0,0,-5,1,-2,-1,-2,-3,-2,0,0,-1,-2,0,0,-1,-8,-6;
/M: SY='G';M=1,-6,3,3,-7,3,0,-4,-1,-5,-4,2,-1,1,-2,1,0,-3,-10,-6;
/M: SY='W';M=-9,-12,-9,-11,1,-11,-4,-8,-5,-3,-6,-6,-8,-7,3,-4,-8,-9,26,0;
/M: SY='W';M=-7,-9,-9,-9,0,-9,-4,-5,-5,-1,-4,-6,-7,-6,2,-3,-6,-6,18,-1;
/M: SY='K';M=-1,-7,0,0,-3,-2,0,-2,2,-3,-1,1,-1,1,2,0,-1,-3,-5,-5;
/M: SY='G';M=2,-3,0,-1,-6,3,-3,-2,-3,-4,-3,0,0,-2,-3,1,0,0,-10,-6;
/M: SY='Q';M=-2,-6,0,0,-3,-3,1,-2,0,-2,-1,0,-2,1,1,-1,-1,-3,-5,-3;
   /I: MI=0; I=-2; MD=0; /M: SY='X'; M=0; D=-2;
/M: SY='T';M=0,-4,-1,-1,-4,0,-2,0,-1,-2,0,0,-1,-1,-1,0,1,0,-7,-5;
/M: SY='T';M=0,-5,0,0,-3,-1,-1,-1,1,-3,-1,1,-1,0,0,1,1,-1,-6,-4;
/M: SY='G';M=0,-5,0,-1,-5,3,-2,-3,-1,-5,-3,0,-1,-1,-1,1,0,-2,-7,-6;
/M: SY='K';M=0,-6,1,1,-5,-1,1,-2,1,-4,-2,2,0,0,0,1,1,-2,-7,-4;
/M: SY='R';M=-1,-6,-1,-1,-5,-3,1,-1,1,-3,-1,0,-1,1,3,-1,-1,-2,-2,-6;
/M: SY='G';M=1,-5,0,0,-6,6,-3,-3,-3,-5,-4,0,-1,-2,-4,1,0,-2,-10,-6;
/M: SY='W';M=-5,-5,-5,-5,2,-6,-2,-2,-4,-1,-3,-3,-6,-5,-3,-3,-4,-4,4,3;
/M: SY='F';M=-3,-5,-6,-6,6,-5,-3,4,-1,3,2,-4,-4,-5,-4,-3,-2,2,-4,3;
/M: SY='P';M=2,-4,-1,-1,-7,-1,0,-3,-2,-4,-3,-1,8,0,0,1,0,-2,-8,-7;
/M: SY='G';M=1,-3,0,0,-4,2,-1,-2,0,-3,-2,0,0,-1,-1,1,1,-1,-6,-5;
/M: SY='N';M=1,-5,2,1,-5,0,1,-2,1,-4,-2,2,0,0,0,1,1,-2,-7,-4;
/M: SY='Y';M=-5,-1,-7,-7,10,-8,-1,-1,-1,-5,-1,-3,-3,-7,-6,-6,-4,-4,-5,0,13;
/M: SY='V';M=0,-3,-3,-5,-2,-2,-3,5,-3,2,2,-2,-2,-3,-4,-1,0,5,-8,-5;
/M: SY='E';M=1,-6,2,3,-6,0,0,-2,1,-4,-2,1,0,2,0,0,0,-3,-8,-6;
/M: SY='P';M=0,-5,-1,-1,-2,-2,-1,-2,-1,-3,-2,0,1,-1,-2,0,-1,-2,-6,-3;
```

Figure 8.5 Example PROSITE profile, showing position-specific scores for insert and match positions. Gap penalties within insert positions are highlighted: in these regions, the values are more permissive, or tolerant, of INDELs by comparison with the high gap penalties for match positions set by the DEFAULT parameter line.

8.3.5 Hidden Markov Models

A variation on the theme of profiles is to encode an alignment in the form of a Hidden Markov Model (HMM). HMMs are probabilistic models consisting of a number of interconnecting states: they are essentially linear chains of match, delete or insert states, which can be used to encode sequence conservation within alignments. A match state is assigned to each conserved column in a sequence alignment; an insert state allows for insertions relative to the match states; and delete states allow match positions to be skipped. Thus, building an HMM from a multiple sequence alignment requires each position within the alignment to be assigned to either match, delete or insert states; this process is illustrated in the linear HMM depicted in Figure 8.6.

HMMs are the basis of the Pfam database. In addition to the HMMs, Pfam also provides the seed alignments used to generate the HMM discriminators, and the final alignments resulting from the iterative sequence gathering process. The alignments are presumed to represent evolutionarily conserved structures that have functional implications. However, unlike the collection of essentially hand-crafted profiles that augment PROSITE patterns, some of the methods that underpin the creation of Pfam are fully automatic. As a result, the iterated group of gathered sequences may become corrupt, i.e., the identified sequences may not all be related. Consequently, the quality of the final alignments, which are not hand-checked, may be quite poor, and any structural and functional implications should be inferred with caution.

We have now examined some of the different pattern recognition methods that underlie the major secondary databases. There are other resources available, but many of these are smaller (e.g., Schultz *et al.'s* SMART domain database) or are less well validated than those chosen for review here (e.g., Smith and Smith's pattern database generated automatically using the PIMA software, or Sonnhammer and Kahn's automatically generated domain database, ProDom, which exploits the DOMAINER program).

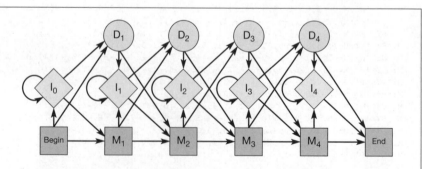

Figure 8.6 Linear Hidden Markov Model, illustrating the possible relationships a sequence may have in matching a multiple alignment; i.e., for each position, or column, of the alignment, a residue may be assigned as a match (M) or an insertion (I), or the position may represent a deletion (D).

Once we understand the differences between the techniques and have appreciated their diagnostic strengths and weaknesses, and provided we keep biological context uppermost in our minds, and appreciate the significance of matching a paralogue, an orthologue, a module, etc., and do not transfer information blindly from a database match to our query, then we can begin to think rationally about how to approach the practical analysis of novel sequences. The following chapter discusses the design of a generalised approach to sequence analysis, with particular reference to an interactive tutorial available on the WWW.

8.4 Summary

- The emphasis of sequence analysis is changing from attempts simply to infer homology, to the more exacting task of recognising orthology. Pattern databases help this process, because they encode different levels of family and species-specific relationships.

- The simplest pattern recognition approach characterises families using single motifs, reducing the sequence data to regular expressions. This is limited because sequences that contain even one residue difference are not matched; conversely, exact matches are not necessarily true, and no context is provided with which to assess their significance.

- Rules are short generic patterns that are not associated with specific families. They are not diagnostic and can only suggest whether a particular functional site might exist.

- Fuzzy regular expressions assign residues to groups with shared biochemical properties. Their tolerance should allow identification of more distant relationships; but the fuzzier the pattern, the greater the chance of making false matches.

- Groups of conserved motifs can be used to create family fingerprints. These are diagnostically potent because of the mutual context provided by motif neighbours, allowing sequence identification even when parts of the signature are absent. Fingerprints do not use scoring matrices to enhance diagnostic power, but most methods do. With all weighting schemes, there is a diagnostic trade-off between the ability to capture true matches and the chance of making more false ones.

- Profiles encapsulate full domain alignments by defining which residues are allowed, which positions are conserved or degenerate, and which positions can tolerate insertions.

- HMMs are linear chains of interconnecting match, delete or insert states that encode full domain alignments. They have similar limitations to other automated iterative approaches, where false matches may corrupt the final discriminators.

8.5 Further reading

Computational biology

BALDI, P. and BRUNAK, S. (1998) *Bioinformatics: The Machine Learning Approach (Adaptive Computation and Machine Learning)*. MIT Press.

DURBIN, R., EDDY, S.R., KROGH, A. and MITCHISON, G. (1998) *Biological Sequence Analysis: Probabilistic Models of Proteins and Nucleic Acids*. Cambridge University Press.

GRIBSKOV, M. and DEVEREUX, J., Eds (1992) *Sequence Analysis Primer*. Oxford University Press.

SCHULZE-KREMER, S. (1995) *Molecular Bioinformatics: Algorithms and Application*. Walter De Gruyter.

SETUBAL, J.C. and MEIDANIS, J. (1997) *Introduction to Computational Molecular Biology*. PWS Publishing Company.

WATERMAN, M.S. (1995) *Introduction to Computational Biology: Maps, Sequences and Genomes*. Chapman and Hall.

Evolution

HUYNEN, M.A. and BORK, P. (1998) Measuring genome evolution. *Proceedings of the National Academy of Sciences, USA*, **95**, 5849–5856.

WEN-HSIUNG, L. and GRAUR, D. (1991) *Fundamentals of Molecular Evolution*. Sinauer Associates, Inc.

Databases and alignment algorithms

SCHULTZ, J., MILPETZ, F., BORK, P. and PONTING, C.P. (1998) SMART, a simple modular architecture research tool: Identification of signaling domains. *Proceedings of the National Academy of Sciences, USA*, **95**, 5857–5864.

SMITH, R.F. and SMITH, T.F. (1992) Pattern-induced multi-sequence alignment (PIMA) algorithm employing secondary structure-dependent gap penalties for use in comparative protein modeling. *Protein Engineering*, **5**, 35–41.

SONNHAMMER, E.L.L. and KAHN, D. (1994) Modular arrangement of proteins as inferred from analysis of homology. *Protein Science*, **3**, 482–492.

CHAPTER NINE

Building a sequence search protocol

9.1 Introduction

This chapter brings together the concepts of primary and secondary database searching within a generalised protocol for sequence analysis. The aim is to gain an understanding of how to interpret results of searching the different data types, and to shed light on the difference between biological and mathematical significance. With this in mind, the chapter shows how to build a search protocol for novel sequences. Application of the protocol is illustrated with reference to an interactive Web tutorial, which seeks to identify an unknown fragment of DNA using the primary, secondary and structure classification databases.

9.2 A practical approach

One of the central goals of bioinformatics is the prediction of protein function, and ultimately of structure, from the linear amino acid sequence. Given a newly determined sequence, we want to know: what is my protein? To what family does it belong? What is its function? And how can we explain its function in structural terms?

Today, although we don't yet have all the answers, we can at least begin to address some of these questions. By searching secondary databases, which house abstractions of functional and structural sites characteristic of particular proteins, we may recognise patterns that allow us to infer relationships with previously characterised families. Similarly, by searching fold libraries, which house templates of known structures, it is possible to recognise a previously characterised fold.

Given the size of current sequence databases, it is increasingly likely that searches with new sequences will uncover homologues; and, with the expansion of sequence pattern and structure template databases, our chances of being able to assign functions and to infer possible fold families are also improving. However, these advances in sequence and fold pattern recognition methods have not yet been matched by similar advances in prediction techniques.

So, if we cannot predict function or structure directly from sequence, but can identify homologues, and recognise sequence and fold patterns that have already been seen, given the bewildering array of databases to search, how do we use this information to build a sensible search protocol for novel sequences?

One practical approach is presented in Figure 9.1. Essentially, we start by checking for identical matches and then move on to search for closely similar sequences in the primary databases. The strategy then involves searching for previously characterised sequence- and, where possible, fold-patterns in a variety of pattern databases. The deciding step is the

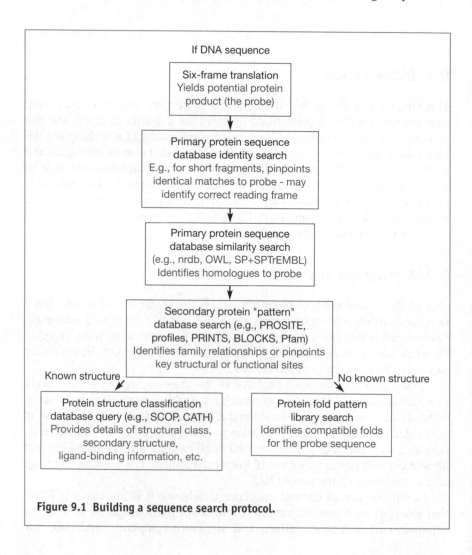

Figure 9.1 Building a sequence search protocol.

family/functional/structural diagnosis. This protocol is described in detail
in the following pages and is accompanied by an interactive WWW tutorial,
known as BioActivity (Attwood, 1997), at:

http://www.bioinf.man.ac.uk/dbbrowser/bioactivity/prefacefrm.html

Before attempting this tutorial, it is important to appreciate that bioinformatics
is a fast-moving field and that Web resources change rapidly. The particular
databases offered within the protocol, and the forms available to search them,
are therefore likely to change with time. The following description of the tutorial can therefore only provide a general guide, with emphasis on how to
assimilate results into a coherent picture, rather than focusing on the minutiae
of how to use particular Web forms for searching particular databases, etc.

9.2.1 Searching the primary databases

To illustrate the steps in building up a search protocol, we will begin by
considering an 'unknown' fragment of DNA. This is fragment G in the Web
tutorial, which can be found in the Materials section at:

http://www.bioinf.man.ac.uk/dbbrowser/bioactivity/nucleicfrm.html

---

```
ccgtactacaactacgctggtgcattcaag
```

Figure 9.2 Example 'unknown' DNA fragment.

The fragment is shown in Figure 9.2. It is translated by inserting it into the
six-frame translator provided on the same page:

http://www.bioinf.man.ac.uk/dbbrowser/bioactivity/nucleicfrm.html

This process yields three forward and three reverse translations, as shown
in Figure 9.3. We now have to discover which of these is the correct reading
frame, i.e. which encodes a real peptide.

Identity searching

The first and fastest test of an unknown protein sequence fragment is to perform an identity search, preferably of a composite sequence database. As
mentioned in Chapter 3, OWL is a composite resource that can be queried
directly by means of its query language. Identity searches, which are suitable
for peptides up to 30 residues in length, are possible via a Web interface; this
provides an easy-to-use form that conveniently shields the user from the syntax
of the query language. An identity search will reveal in a matter of seconds
whether an exact match to the unknown peptide already exists in the database.

In this exercise, each of the translations shown in Figure 9.3 is taken in
turn and used to search the composite database (the 'OWL database search
by sequence' option of the same page:

http://www.bioinf.man.ac.uk/dbbrowser/bioactivity/nucleicfrm.html)

```
                    Forward 0
                          10
                0 PYYNYAGAFK

                    Forward 1
                          10
                0 RTTTTLVHS

                    Forward 2
                          10
                0 VLQLRWCIQ

                    Reverse 0
                          10
                0 LECTSVVVR

                    Reverse 1
                          10
                0 LNAPA!L!Y

                    Reverse 2
                          10
                0 !MHQRSCST
```

Figure 9.3 Forward and reverse translations of the unknown DNA fragment.

In doing this, we find that, with one exception, all of the translations fail to retrieve matches from the database: the query peptide PYYNYAGAFK makes exact matches with two sequences, which are revealed to be transferrins from the African clawed frog, *Xenopus laevis* – Figure 9.4.

Inspection of the ID codes shows that the first match comes from the SWISS-PROT database and the second is a translation from GenBank, which probably contains frameshift errors (hence the duplication of this sequence in OWL – see Section 3.4.2).

```
Matches for SEQ probe PYYNYAGAFK are:

No. of matches = 2

TRFE_XENLA     207     AGIKEHKCSRSNNE PYYNYAGAFK CLQDDQGDVAFVKQ
XLTRSFER       207     AGIKEHKCSRSNNE PYYNYAGAFK CLQDDQGDVAFVKQ

WORKLIST ENTRIES (2):

TRFE_XENLA  TRANSFERRIN PRECURSOR - XENOPUS LAEVIS (AFRICAN CLAWED FROG)
XLTRSFER    X.laevis mRNA for transferrin - African clawed frog.
```

Figure 9.4 Result of an identity search of OWL with a short query sequence. The two matches come from different source databases: TRFE_XENLA is a SWISS-PROT sequence, and XLTRSFER is from GenBank.

If an identity search fails to find a match, all is not lost. The next step is to look for similar sequences, again preferably in a composite database. For best results, it is recommended to perform similarity searches on peptides that are longer than ~30 residues (otherwise, the shorter the peptide, the greater the likelihood of finding chance matches that have no biological relevance).

In our example, although we have identified that our peptide is a transferrin, at this stage we do not know whether a wider set of similar sequences exists in the database. The second stage of the search protocol is thus to perform a similarity search. The most rapid and simple option here is to use one of the many Web interfaces to BLAST (or FastA), the actual choice probably depending on geographical location and Web traffic.

In most applications, where practicable, as much sequence information as possible should be used in a BLAST search (although this can lead to complications in interpreting output from searches with multi-domain or modular proteins). For the purpose of our tutorial, to save network traffic, only the first 180 residues of the transferrin sequence were used: the full sequence (TRFE_XENLA) was retrieved from the primary database using the 'OWL database search by ID code' option of the 'Protein sequence analysis – Primary database searches' page:

http://www.bioinf.man.ac.uk/dbbrowser/bioactivity/proteinfrm.html

and the N-terminal chunk was cut and pasted into the BLAST input facility on the same page (options are provided here to search both the OWL and NRDB composite databases).

The result of searching OWL29.4 is shown in Figure 9.5. There are several important features to note in the BLAST output. First, we are looking for matches that have high scores with correspondingly low probability values. A very low probability indicates that a match is unlikely to have arisen by chance. As the probability values approach unity, they are considered more and more likely to be random matches. The second feature of interest is whether the results show a *cluster* of high scores (with low probabilities) at the top of the list, indicating a likely relationship between the query and the family of sequences in the cluster.

In this example, our transferrin sequence is revealed to belong to an extended family that includes closely related milk lactoferrins, egg white ovotransferrins, blood serotransferrins, membrane-associated melanotransferrins, and saxiphilins. This family is found to cluster towards the top of the hitlist, with high scores and low p-values. Probability values tending to 1.0 indicate random matches and sequences to which the program could not assign a significant score.

Heuristic search tools like BLAST do not always give clear-cut answers. Frequently the program will not be able to assign significant scores to any of its retrieved matches, even if a biologically relevant sequence appears in the hitlist. Such search tools do not have the sensitivity always to fish out the right answer from the sea of sequences in the primary databases; rather, they cast a coarse net, and it is then up to the user to pick out the best prizes from the general catch.

A practical approach

```
BLASTP 1.4.9MP [26-March-1996] [Build 16:16:54 Mar 28 1996]

Reference: Altschul, Stephen F., Warren Gish, Webb Miller, Eugene W.Myers, and
David J.Lipman (1990). Basic local alignment search tool. J.Mol.Biol.215:403-
10.

Query= unknown
         (180 letters)

Database: owl29_4.fasta
         198,742 sequences; 62,935,258 total letters.
Searching.................................................done

                                                          Smallest
                                                            Sum
                                                        High Probability
Sequences producing High-scoring Segment Pairs:         Score P(N)      N

TRFE_XENLA TRANSFERRIN PRECURSOR. - XENOPUS LAEVIS (AFRIC...   960 3.9e-130  1
XLTRSFER XLTRSFER NID: g65158 - African clawed frog.          873 1.5e-116  1
SAX_RANCA SAXIPHILIN PRECURSOR (SAX). - RANA CATESBEIANA...   263 2.8e-60   2
D89084 D89084 NID: g1694683 - Oncorhynchus kisutch cD...      228 7.3e-54   3
TRFE_CHICK OVOTRANSFERRIN PRECURSOR (CONALBUMIN) (ALLERGE...  429 1.4e-53   1
TRF1_SALSA SEROTRANSFERRIN I PRECURSOR (SIDEROPHILIN I). ...  232 7.1e-53   3
NRL_1DOT duck ovotransferrin - duck                           401 1.2e-49   1
TRFL_HUMAN LACTOTRANSFERRIN PRECURSOR (LACTOFERRIN). - HO...  269 6.1e-45   4
NRL_1HSE lactoferrin n-terminal half-molecule mutant H2...    274 3.0e-44   4
NRL_1DSN lactoferrin n-terminal lobe residues 1 333 mut...    268 1.1e-43   4
TRFE_RABIT SEROTRANSFERRIN PRECURSOR (SIDEROPHILIN) (BETA...  277 2.5e-43   4
TRFE_HUMAN SEROTRANSFERRIN PRECURSOR (SIDEROPHILIN) (BETA...  282 2.6e-43   4
NRL_1LFH Lactoferrin (apo form) - human                       274 2.3e-42   4
NRL_1LFI Lactoferrin (copper form) - human                    274 2.3e-42   4
NRL_1LFG Lactoferrin (diferric) - human                       274 2.3e-42   4
TRFE_HORSE SEROTRANSFERRIN PRECURSOR (SIDEROPHILIN) (BETA...  327 2.6e-42   2
NRL_1LCF Lactoferrin (copper and oxalate form) - human        272 4.3e-42   4
TRFL_BOVIN LACTOTRANSFERRIN PRECURSOR (LACTOFERRIN). - BO...  266 1.3e-41   4
TRFL_MOUSE LACTOTRANSFERRIN PRECURSOR (LACTOFERRIN) (FRAG...  267 2.8e-41   3
TRFE_PIG SEROTRANSFERRIN (SIDEROPHILIN) (BETA-1-METAL B...    282 3.6e-41   3
TRFL_PIG LACTOTRANSFERRIN PRECURSOR (LACTOFERRIN). - SU...    187 4.7e-39   5
TRFM_HUMAN MELANOTRANSFERRIN PRECURSOR (MELANOMA-ASSOCIAT...  161 9.9e-33   4
TRF_BLADI TRANSFERRIN PRECURSOR. - BLABERUS DISCOIDALIS ...    105 6.3e-29   4
NRL_1OVB Ovotransferrin (18 kda fragment, domain ii fro...    212 3.9e-24   1
TRF_MANSE TRANSFERRIN PRECURSOR. - MANDUCA SEXTA (TOBACC...     82 2.3e-12   5
S68986 transferrin - flesh fly (Sarcophaga peregrina)          82 2.7e-08   3
S67218 S67218 NID: g456777 - Mus sp. placenta.                 98 5.3e-07   1
B28438 transferrin - mouse (fragment)                          62 0.0058   2
CER01E66 CER01E6 NID: g1082133 - Caenorhabditis elegans.       51 0.70     2
JC2221 major surface glycoprotein 1 - Pneumocystis ca...       50 0.73     3
PAR51C PAR51C NID: g159974 - P.tetraurelia DNA.                53 0.80     3
CELF10G25 CELF10G2 NID: g1458253 - Caenorhabditis elegan...     63 0.95     1
YLJ5_CAEEL HYPOTHETICAL CALCIUM-BINDING PROTEIN C50C3.5 I...    45 0.97     1
S02145 transferrin - horse (fragment)                          52 0.996    1
CEF35G122 CEF35G12 NID: g559423 - Caenorhabditis elegans.      44 0.997    4
S08031 nucleocapsid protein - human coronavirus                61 0.998    1
10KD_VIGUN 10 KD PROTEIN PRECURSOR (CLONE PSAS10). - VIGN...    42 0.999    2
GDIB_MOUSE RAB GDP DISSOCIATION INHIBITOR BETA (RAB GDI B...    42 0.9991   3
PHY4_AVESA PHYTOCHROME A TYPE 4 (AP4). - AVENA SATIVA (OAT).    41 0.9992   3
NCAP_CVH22 NUCLEOCAPSID PROTEIN. - HUMAN CORONAVIRUS (STR...    60 0.9998   1
CELF55C122 CELF55C12 NID: g1086791 - Caenorhabditis elega...    60 0.9999   1
```

**Figure 9.5 Part of a BLAST output from a search of OWL with the query sequence
TRFE_XENLA.** A family of transferrins and related sequences is found to cluster towards the
top of the hitlist. Probability values tending to 1.0 indicate random matches and sequences
to which the program could not assign a significant score.

In these circumstances, where no individual high-scoring sequence, or cluster of sequences, is found, the third feature to consider is whether there are any observable trends in the *type* of sequences matched, i.e., do the annotations suggest that several of these are from a similar family? If there are possible clues in the annotations, the next step is to try to confirm these possibilities both by reciprocal BLAST searches (do retrieved matches identify your sequence in a similarity search?), and by comparing results from searches of the secondary databases.

9.2.2 Searching the secondary databases

Searching PROSITE

For the sake of our tutorial, although a family of sequences was identified by the BLAST search, we will continue on and search the secondary databases in order to discover if our query sequence contains any known characteristic conserved motifs from which we may glean further clues as to its structure or function.

The first secondary database to consider is PROSITE. Within the tutorial, this is accessible for searching via the 'Protein sequence analysis – Secondary database searches' page:

```
Scan of TRFE_XENLA (P20233)

TRANSFERRIN PRECURSOR.
XENOPUS LAEVIS (AFRICAN CLAWED FROG).

[1] PDOC00182 PS00205 TRANSFERRIN_1
Transferrins signature 1

Number of matches: 4
        1     111-120  YYAVAVVKKS
        2     112-120  YAVAVVKKS
        3     443-452  YYAVAIVKKG
        4     444-452  YAVAIVKKG

[2] PDOC00182 PS00206 TRANSFERRIN_2
Transferrins signature 2

Number of matches: 2
        1 211-227  YAGAFKCLQDDQGDVAF
        2 539-554  YSGAFRCLVEKGQVGF

[3] PDOC00182 PS00207 TRANSFERRIN_3
Transferrins signature 3

Number of matches: 2
1     240-270  DYELLCPDNTRKSIKEYKNCNLAKVPAHAVL
        580-610  DFELLCPDGSRAPVTDYKRCNLAEVPAHAVV
```

Figure 9.6 Results returned from a search of PROSITE with the sequence TRFE_XENLA via the ScanProsite WWW interface.

http://www.bioinf.man.ac.uk/dbbrowser/bioactivity/protein1frm. html

The database code (TRFE_XENLA) is simply supplied to the relevant part of the form, and the option to exclude patterns with a high probability of occurrence (i.e., rules) is switched on.

The result, shown in Figure 9.6, indicates that three regular expression patterns have been matched, designated TRANSFERRIN_1, TRANSFER-RIN_2 and TRANSFERRIN_3. Each of the patterns is correctly matched *twice* by the query sequence; judging by the similarity of these matches, this is presumably the result of domain duplication. The first pattern makes an additional match, one residue out of register with the first – this is possible because the pattern tolerates an insertion at the second position of the motif, as denoted by the range 0–1:

```
Y-x(0,1)-[VAS]-V-[IVAC]-[IVA]-[IVA]-[RKH]-[RKS]-[GDENSA]
```

Searching the profile library

The next step in the protocol is to search the ISREC profile library. In addition to the profiles that have already been incorporated into the main body of PROSITE, the Web server offers a range of pre-release profiles that have not yet been sufficiently documented for release through PROSITE. Searching the complete collection of profiles is achieved, once again, by simply supplying the database code to the Web form, remembering to change the format button from the default (plain text) to accept a SWISS-PROT ID:

http://www.bioinf.man.ac.uk/dbbrowser/bioactivity/protein1frm.html

```
Sequence: TRFE_XENLA was searched against PROSITE pre-release profiles
library, the sensitivity was set to default

In addition to the raw score, the normalized NScore is reported. Note
that if you searched the Pfam bitscaled database, the normalized score
shown is identical to the original HMMER bitscore.

Hits coming from PROSITE Profiles are color-coded in red
Hits coming from PROSITE Patterns in magenta
Hits coming from Pfam in blue
Hits coming from Gribskov collection in green

Significant matches are labeled with a red "!" in the first column.
If no match is found, the list below remains empty
----------------------------------------------------------------------


normalized raw        from -   to Profile | Description

For a graphical display of the hitlist, use the SEView Applet


----------------------------------------------------------------------
```

Figure 9.7 Results of searching the PROSITE and pre-release profile library with sequence TRFE_XENLA. As indicated, when there are no matches to a profile, the results box remains empty, as witnessed here.

The result of searching the profiles collection is illustrated in Figure 9.7. No matches were found in this resource; but, in fact, later inspection reveals that there is no transferrin profile in this version of the library, so we could not have expected a positive result anyway.

Searching Pfam

Another important resource to search is the Pfam collection of Hidden Markov Models. Searching is achieved via a Web interface that requires the query sequence to be supplied to a text box:

http://www.bioinf.man.ac.uk/dbbrowser/bioactivity/protein1frm.html

but the sequence must be in FastA format, which means that the query must be preceded by the > symbol and a suitable sequence name, as shown in Figure 9.8.

```
>TransferrinQuery
M D F S L R V A L C L S M L A L C L A I Q K E K Q V R W C V
K S N S E L K K C K D L V D T C K N K E I K L S C V E K S N
T D E C S T A I Q E D H A D A I C V D G G D V Y K G S L Q P
Y N L K P I M A E N Y G S H T E T D T C Y Y A V A V V K K S
S K F T F D E L K D K K S C H T G I G K T A G W N I I I G L
L L E R K L L K W A G P D S E T W R N A V S K F F K A S C V
```

Figure 9.8 Example of a query sequence in FastA format.

The result of the Pfam search is shown in Figure 9.9. This analysis identifies three regions with significant alignments, the first spanning the first domain revealed by the PROSITE search, the second and third regions lying within the second domain.

```
Sequence TransferrinQuery - Domain organisation
```

Domain	Start	End	Bits	Alignment
transferrin	26	341	669.51	Align
transferrin	354	484	123.16	Align
transferrin	493	686	186.64	Align

Figure 9.9 Results returned from a search of Pfam with the query sequence TRFE_XENLA.

Searching PRINTS

Another key secondary resource is PRINTS, which provides a bridge between single-motif search methods, such as the one used to compile PROSITE, and domain-alignment/profile methods, such as those embodied in the profile library and Pfam.

PRINTS is accessible for searching via the 'Protein sequence analysis – Protein fingerprinting' page:

http://www.bioinf.man.ac.uk/dbbrowser/bioactivity/protein2frm.html

The results of searching PRINTS are illustrated in Figure 9.10. The output is divided into distinct sections: first, the program offers an intelligent 'guess' based on the occurrence of the highest-scoring complete or partial fingerprint match, or matches; it then provides an expanded table that shows the top 10 best-scoring matches – clearly, these include the intelligent result from the previous table, but the additional matches are provided to highlight why the best guess was chosen, and to allow a different choice, if the guess is considered either to be wrong or to have missed something; the remaining sections of output provide more of the raw data, again allowing the user to search for anything that might have been missed.

In the result shown in Figure 9.10, we see that the intelligent guess is a

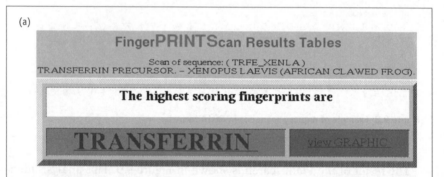

(a)

FingerPRINTScan Results Tables

Scan of sequence: (TRFE_XENLA)
TRANSFERRIN PRECURSOR. – XENOPUS LAEVIS (AFRICAN CLAWED FROG).

The highest scoring fingerprints are

TRANSFERRIN   view GRAPHIC

(b)

Simple format						
Fingerprint	**No.Mots**	**Sum**	**Ave**	**Score**	**GRAPHScan**	
TRANSFERRIN	5 of 5	347.16	69.43	6902.71	IIIII	Graphic
CHYMOTRYPSIN	3 of 3	50.89	16.96	746.09	iii	Graphic
ELONGATNFCT	4 of 5	69.36	17.34	725.86	.iiii	Graphic
P450	3 of 5	60.74	20.25	723.32	.i.ii	Graphic
BASICPTASE	3 of 3	50.61	16.87	711.55	iii	Graphic
LYZLACT	3 of 6	64.48	21.49	689.02	.i.i.i	Graphic
SH2DOMAIN	3 of 5	53.25	17.75	653.13	.i.ii	Graphic
CYTCHRMECIAB	3 of 6	68.27	22.76	646.03	i.i..i	Graphic
ALPHAHAEM	3 of 5	52.42	17.47	631.56	i.ii.	Graphic
BETALLERGEN	3 of 7	57.56	19.19	616.83	.ii...i	Graphic

Figure 9.10 Results returned from a search of PRINTS with the query sequence TRFE_XENLA: (a) the program's 'best guess'; (b) the extended hitlist from which the 'guess' was deduced.

match with the TRANSFERRIN fingerprint. Examination of the expanded table highlights the quality of this match compared with all other matches, in terms of the number of motifs matched, the average scores, and so on. It is clear from this extended hitlist that the program's guess was a good one and that transferrin is indeed the best match.

A particularly valuable aspect of this software is the facility to visualise individual fingerprint matches by clicking on the 'graphic' box on the right-hand side of the table. The result is illustrated in Figure 9.11: in the graph, the x-axis denotes the sequence, and the y-axis, which is divided into the number of motifs that make up the fingerprint, denotes percent identity of the match (0–100 per motif). Filled blocks mark the positions of the highest-scoring matches between particular motifs and the query sequence. In the example shown, the five motifs of the TRANSFERRIN fingerprint are scanned along the TRFE_XENLA query sequence, each motif making a strong match, as denoted by the series of blocks matching in order from the N- to the C-terminus.

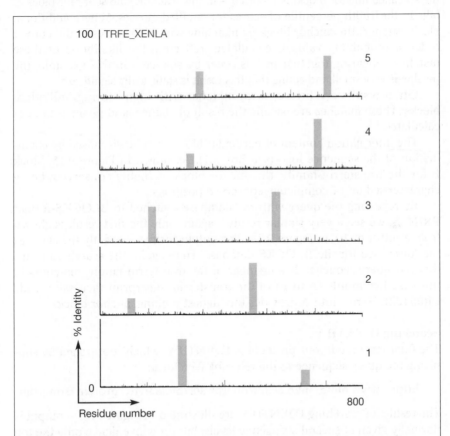

Figure 9.11 Graphical result returned from a search of the query sequence TRFE_XENLA with the TRANSFERRIN fingerprint. Filled blocks appearing in the correct order along the query sequence mark a strong match between the query and the given fingerprint.

Searching the BLOCKS databases

The next secondary resources to be searched in this protocol are the BLOCKS databases, derived from PROSITE and PRINTS. If results matched in PROSITE and/or PRINTS are true-positive, then we would expect these to be confirmed by the BLOCKS search results.

Within the tutorial, the BLOCKS databases are searched by supplying the query sequence to the input box of the relevant Web form:

http://www.bioinf.man.ac.uk/dbbrowser/bioactivity/protein1frm.html

remembering in each case to switch to the required database. The results of these database searches are illustrated in Figures 9.12 and 9.14.

At first sight, the output may appear daunting, but one or two pointers make it easier to understand. First, it is clear from the Description line that the sequence has been identified as a transferrin. The accession codes (BL00205) in the Block column indicate that there are five motifs, labelled A to E; matches to these motifs are ranked according to score. The 'rank' of the best-scoring block, the so-called anchor block, is reported – in this example, the score for block B falls in the 100th percentile of scores for shuffled queries. Where additional blocks support the anchor block by matching with high scores in the correct order, a probability value is calculated, reflecting the likelihood of these matches appearing together in this order by chance – in this example, the p-value is very small, indicating that this result is statistically significant.

Often results are littered with matches with high-scoring individual blocks. These matches are usually the result of chance, and p-values are not calculated.

The information content of particular blocks can be visualised by examination of the sequence logo (see Box 9.1). As shown in Figure 9.13, block B for the transferrin family, the anchor block, is highly conserved, being characterised by 14 completely conserved positions.

By repeating the query with the database switched to BLOCKS-format PRINTS, we see a very similar result – again, only the first result is shown from a hitlist of 10, in Figure 9.14. The output format is exactly the same as that described for the BLOCKS database. Here again, the search confirms that our query sequence is a member of the transferrin family, matching in this case five motifs (A to E) of the transferrin fingerprint (accession code PR00422). Here, motif A provides the highest ranking, anchor block.

Searching IDENTIFY

The final resource in our protocol is IDENTIFY, which is searched by supplying the query sequence to the relevant Web form:

http://www.bioinf.man.ac.uk/dbbrowser/bioactivity/protein1frm.html

The results of searching IDENTIFY are illustrated in Figure 9.15 – output is normally given at several stringency levels, but for convenience only the top level is shown here.

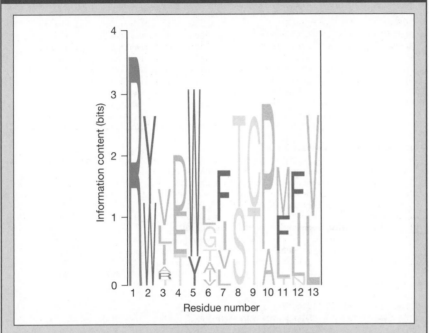

A sequence logo is a graphical display of a multiple alignment consisting of colour-coded stacks of letters representing amino acids at successive positions. The height of a given letter increases with increasing frequency of the amino acid, and its height increases with increasing conservation of the aligned position; hence, letters in stacks with single residues (i.e., representing conserved positions) are taller than those in stacks with multiple residues (i.e., where there is more variation).

Within stacks, the most frequently occurring residues are not only taller, but also occupy higher positions in the stack, so that the most prominent residue at the top is the one predicted to be the most likely to occur at that position. To address the problem of sequence redundancy within a block, which strongly biases residue frequencies, sequence weights are calculated using a position-specific scoring matrix (PSSM). This reduces the tendency for over-represented sequences to dominate stacks, and increases the representation of rare amino acids relative to common ones. Thus, for example, at a position within a block where six leucines and two methionines are observed, a sequence-weighted logo will nevertheless place methionine above leucine, because the leucines are relatively over-represented.

Clearly, this resource also confirms that the query sequence belongs to the transferrin family. The first match corresponds to PROSITE pattern TRANSFERRIN_1, and the following five to PRINTS motifs TRANSFERRIN1–5.

```
1.-----------------------------------------------------------------------
Block     Rank Frame Score Strength    Location (aa) Description
BL00205A    7    0   1596  1778          128-   151 Transferrins proteins.
BL00205A    8    0   1565  1778          460-   483 Transferrins proteins.
BL00205B    1    0   1743  1815          206-   233 Transferrins proteins.
BL00205B   10    0   1148  1815          534-   561 Transferrins proteins.
BL00205C    2    0   1709  1855          241-   272 Transferrins proteins.
BL00205C    6    0   1606  1855          581-   612 Transferrins proteins.
BL00205D    3    0   1686  1875          456-   483 Transferrins proteins.
BL00205D    5    0   1629  1875          124-   151 Transferrins proteins.
BL00205E    4    0   1675  1781          534-   548 Transferrins proteins.
BL00205E    9    0   1360  1781          206-   220 Transferrins proteins.


1743=100.00th percentile of anchor block scores for shuffled queries
P<8.4e-15 for BL00205C BL00205D BL00205E BL00205A in support of BL00205B
Maximum number of repeats (from Prosite MAX-REPEAT) = 2
1 non-overlapping repeats in support of BL00205A
1 non-overlapping repeats in support of BL00205B
1 non-overlapping repeats in support of BL00205C
1 non-overlapping repeats in support of BL00205D
1 non-overlapping repeats in support of BL00205E
                        |--- 180 amino acids---|
   BL00205 AAA:::::::.BBBB:.CCCC:::::::::::::::::::::::::..DDDD:::::::EE
   Unknown <AAA:::::::BBBB:CCCC:::::::::::::::::::::::::::::DDDD:::::::EE
   Unknown <                                   AAA        BBB
   Unknown DDDD       EE


BL00205A   <->A    (110,130):127
TRF1_SALSA 121     LRGKKSCHTGLGKSAGWNIPIGTL
                   |  |||||||  || |||||  ||| |
Unknown    128     LkdKKSCHTGiGKTAGWNIiIGLL
           460     LRGvKtCHTavGRTAGWNIPvGLi


BL00205B   A<->B   (47,54):54
TRFE_HORSE 204     EPYFGYSGAFKCLADGAGDVAFVKHSTV
                   |||  |  |||||| |   ||||||| |||
Unknown    206     EPYYnYAGAFKCLQDDqGDVAFVKQSTV
           534     EaYYGYSGAFrCLvEkgqvgfakhtTvf


BL00205C   B<->C   (7,12):7
TRF1_SALSA 231     YELLCKDGTRASIDSYKTCHLARVPAHAVVSR
                   |||||  |  || ||   || |  || |||||||  |
Unknown    241     YELLCPDNTRKSIkEYKNCnLAkVPAHAVltR
           581     fELLCPDGSRAPVtDYKrCnLAeVPAHAVVtl


BL00205D   C<->D   (178,190):183
TRFE_HUMAN 459     TWDNLKGKKSCHTAVGRTAGWNIPMGLL
                   | || | | |||||||||||||||| ||
Unknown    456     sWsNLrGvKtCHTAVGRTAGWNIPvGLI
           124     TfDeLKDKKSCHTGiGkTAGWNIiiGLL


BL00205E   D<->E   (42,50):50
TRF1_SALSA 518     EQYYGYTGAFRCLVE
                   | |||| |||||||||
Unknown    534     EaYYGYsGAFRCLVE
           206     EpYYnYaGAFkCLqd
```

Figure 9.12 Results returned from a BLOCKS search with the query sequence TRFE_XENLA.

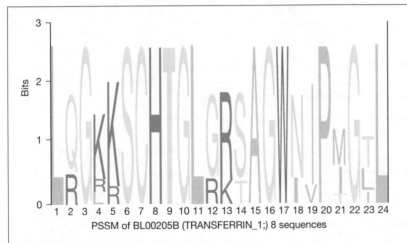

PSSM of BL00205B (TRANSFERRIN_1;) 8 sequences

Figure 9.13 Sequence logo for transferrin block B, highlighting the most conserved positions within the block.

```
1.-----------------------------------------------------------------------
Block     Rank Frame Score Strength   Location (aa) Description
PR00422A    1    0   1701  1587        208-   229  TRANSFERRIN SIGNATURE
PR00422A    8    0   1198  1587        536-   557  TRANSFERRIN SIGNATURE
PR00422B    4    0   1563  1362        398-   417  TRANSFERRIN SIGNATURE
PR00422B   13    0   1109  1362         73-    92  TRANSFERRIN SIGNATURE
PR00422C    2    0   1658  1566        460-   480  TRANSFERRIN SIGNATURE
PR00422C    5    0   1538  1566        128-   148  TRANSFERRIN SIGNATURE
PR00422D    3    0   1597  1388        572-   590  TRANSFERRIN SIGNATURE
PR00422D   10    0   1118  1388        232-   250  TRANSFERRIN SIGNATURE
PR00422E    6    0   1429  1331        599-   615  TRANSFERRIN SIGNATURE
PR00422E    7    0   1275  1331        259-   275  TRANSFERRIN SIGNATURE
1701=100.00th percentile of anchor block scores for shuffled queries
P<5.1e-13 for PR00422C PR00422D PR00422B PR00422E in support of PR00422A
                      |--- 188 amino acids---|
  PR00422 AAA::::::::::::::::::::::...BBB::::..CCC:::::::::::....DDD:EE
  Unknown AAA::::::::::::::::::::::::BBB::::::CCC::::::::::::DDD:EE
  Unknown <                                        AAA
  Unknown <  DDD EE

PR00422A   <->A   (149,352):207
TRFE_XENLA 208   YYNYAGAFKCLQDDQGDVAFVK
                 ||||||||||||||||||||||
Unknown    208   YYNYAGAFKCLQDDQGDVAFVK

PR00422B   A<->B  (163,186):168
TRFE_XENLA 398   ADAVTLDGGYMYTAGLCGLV
                 |||||||||||||||||||||
Unknown    398   ADAVTLDGGYMYTAGLCGLV

PR00422C   B<->C  (32,45):42
TRFE_XENLA 460   LRGVKTCHTAVGRTAGWNIPV
                 |||||||||||||||||||||
Unknown    460   LRGVKTCHTAVGRTAGWNIPV

PR00422D   C<->D  (83,111):91         PR00422E D<->E (8,9):8
TRFE_XENLA 572   WAKDLKSEDFELLCPDGSR   TRFE_XENLA 599 CNLAEVPAHAVVTLPDK
                 ||||||||||||||||||               |||||||||||||||||
Unknown    572   WAKDLKSEDFELLCPDGSR   Unknown    599 CNLAEVPAHAVVTLPDK
```

Figure 9.14 Results from a search of BLOCKS-format PRINTS with the query sequence TRFE_XENLA.

```
At a stringency of at least one in 1010 (no false positives expected)

Name          Description  Motif

TRANSFERRIN_1Transferrins y[fy].y[as]gaf.cl.[de]..gdvaf[iv][kr]..[ast][iv]
             positions
             207-233
...AGIKEHKCSRSNNEPYYNYAGAFKCLQDDQGDVAFVKQSTVPEEFHKDYELLCPDN...

TRANSFERRIN   TRANSFERRINS  [fy]....ga[fly].cl....g.va[fwy]
             SIGNATURE
             positions
             207-227        ...AGIKEHKCSRSNNEPYYNYAGAFKCLQDDQGDVAFVKQSTVPEEFHKDYE...

TRANSFERRIN  TRANSFERRIN   ada[ilmv].[ilv][de]....y.a..cgl[ilv]
             SIGNATURE
             positions
             397-417        ...ASTAEECIVQILKGDADAVTLDGGYMYTAGLCGLVPVMGEYYDQDDLTPC...

TRANSFERRIN  TRANSFERRIN   [ilv][ekqr]..[kr][st]ch[ast].[filvy]...agw.[iv]p[ilmv]
             SIGNATURE
             positions
             459-480        ...AVAIVKKGTQVSWSNLRGVKTCHTAVGRTAGWNIPVGLITSETANCDFASY...

TRANSFERRIN  TRANSFERRIN   w[ast]..1...[dn][fy].[ilv]lc.[dn]
             SIGNATURE
             positions
             571-587        ...HTTVFENTDGKNPAGWAKDLKSEDFELLCPDGSRAPVTDYKRCNLA...

TRANSFERRIN  TRANSFERRIN   c.l[as].[iv]p...[iv]vt..[de]
             SIGNATURE
             positions
             598-614        ...LCPDGSRAPVTDYKRCNLAEVPAHAVVTLPDKREQVAKIVVNQQSL...
```

Figure 9.15 Results returned from a search of IDENTIFY with the sequence TRFE_XENLA.

9.3 When to believe a result

The results of these primary and secondary database searches are sum-marised in Table 9.1. Taken together, the emerging picture is that our sequence evidently belongs to the transferrin family. This diagnosis is sup-ported by all but one method, where, at the time of the search, there was clearly no information available for the family.

As an example of the application of this type of search protocol, the transferrins are straightforward; the diagnosis is clear. The real need for such a strategy arises when query sequences do not have obvious homo-logues in the database, i.e., do not return positive identifications following sequence identity or similarity searches of the primary databases. Under these circumstances, composing a diagnostic picture from searches of a range of secondary databases is essential. If two or more of these make twi-light matches, this builds confidence in making a diagnosis.

9.4 Structural and functional interpretation

Application of this sequence search protocol to our unknown 30-base frag-ment of DNA, at a superficial level, has told us little more than that the

Table 9.1 Summary of search protocol results. All but one method returned a positive diagnosis for the query sequence. Asterisked resources are those that provide family annotations.

Method	Diagnosis
OWL ID SEARCH	TRANSFERRIN SEQUENCE
BLAST	TRANSFERRIN FAMILY
PROSITE*	TRANSFERRIN FAMILY
PROFILES	NO MATCH
PFAM	TRANSFERRIN
PRINTS*	TRANSFERRIN FAMILY
BLOCKS	TRANSFERRIN FAMILY
BLOCKS-PRINTS	TRANSFERRIN FAMILY
IDENTIFY	TRANSFERRIN FAMILY

protein product is a member of the transferrin family. Thus far, we have only scratched the surface; we have learned nothing of what a transferrin protein might do, or what it might look like.

To glean this kind of information, it is necessary to delve into the annotations or documentations provided by some of the secondary databases. PROSITE and PRINTS are manually annotated resources – they alone provide family descriptions, and, where possible, details of the significance and locations of the conserved motifs encoded in the databases. So, what more can we learn about transferrins?

9.4.1 What do transferrins do?

Inspection of the PROSITE result allows us to follow links back to the family documentation file (PDOC00182). Here, we learn that transferrins are eukaryotic iron-binding glycoproteins that function to control the level of free iron in biological fluids. The proteins, which bind two atoms of ferric iron in association with an anion (usually bicarbonate), transport iron from sites of absorption and haem degradation to those of storage and utilisation.

Transferrins are believed to have evolved through duplication of a 340-residue iron-binding domain, in which each iron atom is bound by four conserved residues: an aspartic acid, two tyrosines and a histidine. The cysteines in both domains form intra-domain disulphide bonds. Each of the three PROSITE patterns includes an iron-binding ligand: a tyrosine in the first two patterns, and a histidine in the third.

Following the link to the PRINTS documentation file (PR00422), we find similar information, and we learn that the first motif of the fingerprint has been drawn from the same region of the alignment as has PROSITE pattern TRANSFERRIN_2, which encodes the conserved iron-binding tyrosine and a cysteine residue involved in disulphide bond formation.

9.4.2 What do transferrins look like?

A further important feature emerges from the PROSITE and PRINTS transferrin entries – each provides cross-references to other biological databases,
which in this instance indicate that co-ordinates of the 3D structure are
available (e.g., PDB file 1tfd). We can find out more about the structure,
either by following the links embedded in the PROSITE and PRINTS transferrin entries, or by supplying a relevant PDB code to the query forms of
the structure classification resources (such as SCOP and CATH).

SCOP

In this tutorial, SCOP is accessible for searching via the 'Protein structure
analysis – Structure classification resources' page:

http://www.bioinf.man.ac.uk/dbbrowser/bioactivity/structurefrm.html

The relevant PDB code is simply supplied to the input box of the Web form.
The results of querying SCOP with code 1tfd are illustrated in Figure 9.16.

Protein: Transferrin from rabbit (Oryctolagus cuniculus)

Lineage:

1.**Root:** scop
2.**Class:** Alpha and beta (a/b)
  Mainly parallel beta sheets (beta-alpha-beta units)
3.**Fold:** Periplasmic binding protein-like II
  consists of two similar intertwined domain with 3 layers (a/b/a) each: duplication
  mixed beta-sheet of 5 strands, order 21354; strand 5 is antiparallel to the rest
4.**Superfamily:** Periplasmic binding protein-like II
  Similar in architecture to the superfamily I but partly differs in topology
5.**Family:** Transferrin
  further duplication: composed of two two-domain lobes
6.**Protein:** Transferrin
  N-terminal lobe
7.**Species:** rabbit (Oryctolagus cuniculus)

PDB Entry Domains:

1.1tfd
  complex with carbonate --; fe(iii) ion

Figure 9.16   Query results from the SCOP database for PDB code 1tfd.

The results returned from this query reveal that the structure whose co-
ordinates are available is rabbit transferrin, a two-domain α–β protein, with
each domain forming a three-layer ($\alpha\beta\alpha$) sandwich structure.

CATH

The CATH resource is queried by supplying the desired PDB code to the
relevant form on the same Web page:

The results of querying CATH with code 1tfd are illustrated in Figure 9.17. The database splits the sequence into two domains, which are assigned CATH numbers 3.40.400.10 and 3.40.190.10 respectively, indicating that both folds are three-layer (αβα) sandwiches from the α–β class, which is consistent with the information gathered from SCOP.

Name: Transferrin (n-terminal half-molecule)
Source: Rabbit (Oryctolagus cuniculus) serum

Summary:

PDB Code	Chain	Status
1tfd	–	In CATH

2 assigned domains

Domain 1: residues 6 to 84, 250 to 303
Goto CATH entry

Class	3	Alpha Beta
Architecture	40	3-Layer(aba) Sandwich
Topology	400	Phosphate-Binding Protein, domain 1
Homologous superfamily	10	1lcf domain 1

Domain 2: residues 85 to 249
Goto CATH entry

Class	3	Alpha Beta
Architecture	40	3-Layer(aba) Sandwich
Topology	190	D-Maltodextrin-Binding Protein, domain 2
Homologous superfamily	10	1dmb domain 2

Figure 9.17 Query results from the CATH database for PDB code 1tfd.

PDBsum

Clicking on the hyperlinked PDB code in the CATH summary takes us to the PDBsum resource, a Web-based collection of information for all PDB structures. Here, we are able to view pictures of the overall fold and secondary structure of the molecule, together with images of the carbonate ligand and of its interactions with the protein; for example, see Figures 9.18 and 9.19.

Using this pictorial information, we can begin to rationalise the results of our secondary database searches in terms of structural and functional features of the 3D molecule, essentially by superposing the motifs matched in PROSITE, PRINTS and BLOCKS onto the sequence. Doing this, we discover that the three PROSITE patterns were drawn from regions in the first domain that are clearly duplicated in the second. Looking at the PRINTS motifs, however, we find that only motif A lies in the first domain, motifs B to E falling in different

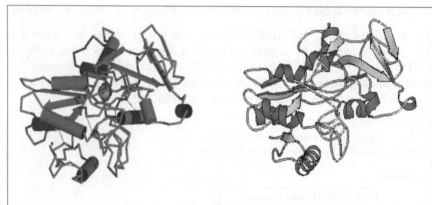

Figure 9.18 Part of the PDBsum entry for PDB code 1tfd, showing different representations of the overall protein fold.

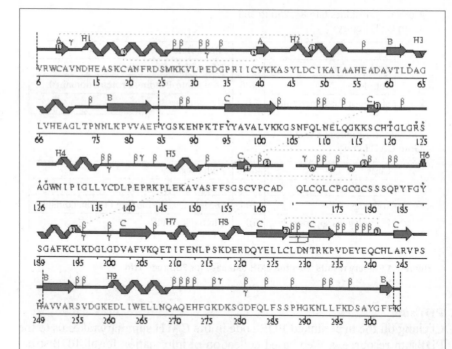

Figure 9.19 Part of the PDBsum entry for PDB code 1tfd, showing the secondary structures of the protein mapped onto the sequence.

regions of the second domain. For BLOCKS, we find that blocks A, B and C fall within the first domain, while blocks D and E are effectively the analogues of blocks A and B in the second domain. This helps us to understand another feature of the BLOCKS search output, in which strings of alphabetic characters (corresponding to the blocks) are depicted along a ruler, which represents the sequence (see Figures 9.12 and 9.14). It is immediately apparent from this sketch that blocks A, B and C cluster towards the N-terminus of the sequence, and blocks D and E towards the C-terminus, and moreover that there is a correspondence both between blocks D and A and between blocks E and B. Regions corresponding to the locations of the motifs are highlighted in Figure 9.20.

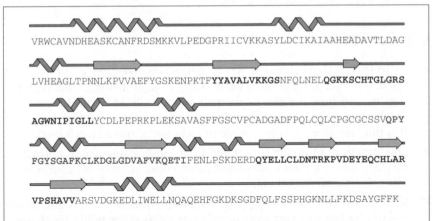

VRWCAVNDHEASKCANFRDSMKKVLPEDGPRIICVKKASYLDCIKAIAAHEADAVTLDAG

LVHEAGLTPNNLKPVVAEFYGSKENPKTF**YYAVALVKKGS**NFQLNEL**QGKKSCHTGLGRS**

AGWNIPIGLLYCDLPEPRKPLEKSAVASFFGSCVPCADGADFPQLCQLCPGCGCSSV**QPY**

FGYSGAFKCLKDGLGDVAFVKQETIFENLPSKDERD**QYELLCLDNTRKPVDEYEQCHLAR**

VPSHAVVARSVDGKEDLIWELLNQAQEHFGKDKSGDFQLFSSPHGKNLLFKDSAYGFFK

Figure 9.20 Illustration of regions of sequence from which motifs have been drawn to develop characteristic signatures for the transferrins. The main secondary structures are shown: grey arrows denote strands; blue ribbons represent helices. Regions marked in bold show where motifs from most of the secondary databases lie, and can be seen to correspond with conserved secondary structures. The region marked blue is the most highly conserved, and spans the helix and strand encoded by all three secondary databases.

While there is clearly some overlap between the motifs identified within the different databases, there is not an exact correspondence. The only region of the sequence at which all methods converge is that encoded by PROSITE pattern TRANSFERRIN_2 (PS00206), PRINTS motif PR00422A and block BL00205B (recall that these were the highest-ranking, anchor blocks in the BLOCKS database searches). This is a highly conserved region and comparison with the structure reveals that it spans part of a β-strand and α-helix 6, which supports both a conserved iron-binding tyrosine residue and a cysteine involved in disulphide bond formation. PROSITE pattern TRANSFERRIN_1 (PS00205) also encodes a β-strand, and four strands are spanned by pattern TRANSFERRIN_3 (PS00207) and block BL00205C (the first of these strands falls within PRINTS motif PR00422D and the second two within motif PR00422E). Finally, both block BL00205A and motif PR00422C span a region that includes a short β-strand and α-helix 4.

That there are differences between the results is not surprising; as we have seen, the analysis methods that underlie the different databases are themselves different, those used to construct PROSITE and PRINTS being essentially manual, those used to derive BLOCKS being fully automatic. In the final analysis, what is important is to achieve a general consistency between results, and to exploit all available annotations to yield the most detailed structural, functional and evolutionary picture possible for any given query.

So far, our investigation has dealt only with the use of static Web forms as an interface to searching various primary and secondary databases, and to querying the structure classification resources. Where 3D co-ordinates are available, as in this example, we can take the analysis a step further using software packages that allow visualisation and interactive manipulation of sequence and structural information. Examples of such packages form the subject of the following chapter.

9.5 Summary

- In trying to characterise a newly determined sequence, we want to know what the protein is, to what family it might belong, what its function is, and how we can explain its function in structural terms. No database or software yet exists that can answer these questions directly. It is thus sensible to build a range of techniques into a search protocol.

- One practical approach is outlined in an on-line interactive tutorial at: http://www.bioinf.man.ac.uk/dbbrowser/bioactivity/prefacefrm.html. The principal steps include:

- Identity searching of a composite database – this is the first and fastest test of whether the exact sequence already exists in the public databases.

- Similarity searching – this will show whether the query belongs to an extended family. For example, in BLAST output, look for matches with high scores and low p-values, consider also whether there are clusters of high scores at the top of the hitlist, otherwise look for trends in the type of sequences matched.

- Pattern database searching – this will indicate whether the query contains any known characteristic motifs that may be suggestive of particular aspects of its structure or function. There are several major resources to search; detailed family information may be gleaned by delving into the annotations provided by PROSITE and PRINTS.

- Fold classification database querying – once a consensus diagnosis has been reached, further information might be accessible (i.e., if a structure is known) by interrogating the fold class databases, or by examining the summary information provided by PDBsum.

- Only by using a range of tools and databases can you gain the most from your sequence analysis, because none of the databases is complete, and none of the search methods is infallible. By marrying results together, like pieces in a jigsaw, a more complete structural, functional and evolutionary picture of your protein should begin to emerge.

9.6 Further reading

ATTWOOD, T.K. (1997) BioActivity: An interactive bioinformatics practical on the WWW. *Life Sciences Educational Computing*, 8(3), 10–13.

BAXEVANIS, A. and OUELLETTE, B.F.F., Eds (1998) *Bioinformatics: A Practical Guide to the Analysis of Genes and Proteins.* John Wiley and Sons.

SALTER, H. (1998) Teaching bioinformatics. *Biochemical Education*, **26**, 3–10.

CHAPTER TEN

Analysis packages

10.1 Introduction

In previous chapters, we have discussed the seemingly bewildering array of biological databases, the algorithms that have been devised to search them, and the kind of analysis protocol that may be pursued via a simple Web interface. This chapter deals briefly with stand-alone analysis packages, touching on the types of facilities they offer and why there was a need to create them. Popular packages from both the commercial sector and the public domain are outlined (e.g., GCG, Staden), and the latest developments on the WWW are highlighted (e.g., CINEMA). Issues relating to software development and licensing are mentioned from both academic and commercial perspectives.

10.2 What's in an analysis package?

In Chapter 9, we saw that in the course of analysing a protein sequence, several search methods need to be applied. But homology searching is only one aspect of the analysis process. Numerous other research tools are also available, including **hydropathy profiles** for the detection of possible **transmembrane domains** and/or hydrophobic protein cores (see Figure 10.1 and Box 10.1); **helical wheels** to identify putative **amphipathic helices** (see Box 10.2); sequence alignment and phylogenetic tree tools for charting evolutionary relationships; secondary structure prediction plots for locating α-helices and β-strands; and so on.

Because of the need to employ a range of techniques for effective sequence analysis, software packages have been developed to bring a variety of these methods together under a single umbrella, obviating the need to use different tools with different interfaces, with different input requirements and different output formats. We will start by considering commercial databases

Analysis packages

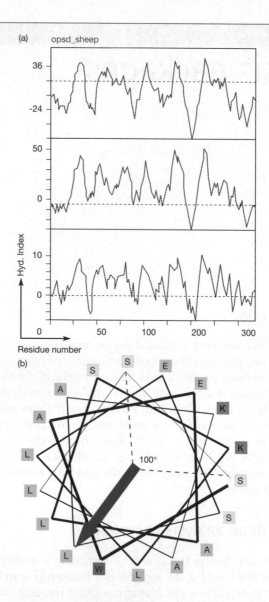

Figure 10.1 Illustration of different tools used in the analysis of protein sequences. (a) Hydropathy profiles, for pinpointing potential membrane-spanning regions or hydrophobic protein cores – here, three different scales are shown (Eisenberg, Kyte and Doolittle, and a 'transmembrane' index). Peaks penetrating above the dotted lines indicate possible transmembrane (TM) domains. Clearly, there is overall similarity between the graphs, but there are differences in detail regarding the number and extent of the potential hydrophobic regions. Where there is a consensus between the methods, there is a greater chance that the predictions are reliable. In this case, for example, there is a measure of confidence that six TM regions might exist, but considerable ambiguity regarding the presence of a possible seventh domain. (b) Helical wheel showing the sequence of an amphipathic α-helix. The separation of hydrophobic and charged residues is clear; the direction of the hydrophobic moment corresponding to this is shown by the arrow.

Hydrophobicity (or hydropathy) is an essential driving force in protein folding. The hydrophobic character of amino acid residues is complex, and no single hydrophobic parameter can represent the complete range of amino acid behaviour. Depending on the context in which hydrophobicity is being considered, the property may be defined in several ways: e.g., as a tendency to exhibit less affinity for water, it may be estimated from physicochemical measurements on particular amino acids; as a tendency to occupy the core of proteins, it may be estimated from statistical properties of residues in 3D structures, where it carries more complex information than a single physical or chemical property. Thus, hydropathy scales broadly fall into two classes, based on (i) experimental measurements of chemical behaviour or amino acid physicochemical properties, such as solubility in water, chromatographic migration, etc.; and (ii) environmental characteristics of protein residues, such as solvent accessibility, propensity to occupy the protein interior, etc.

Accordingly, there are many different hydropathic rankings for the amino acids (several of these have been reviewed by Trinquier and Sanejouand, 1998). At the most basic level, 'internal, external, ambivalent' (i = {FILMV}, e = {DEHKNQR} and a = {ACGPSTWY}) provides a good alphabet for distinguishing hydrophobicity. More sensitive scales assign values to individual amino acids, and special scales have been devised to detect transmembrane helices (these are based on amino acid distributions particular to membrane proteins). Some typical rankings are noted below:

Scale name	Residue ranking
Zimmerman	LYFPIVKCMHRAEDTWSGNQ
von Heinje	FILVWAMGTSYQCNPHKEDR
Efremov	IFLCMVWYHAGTNRDQSPEK
Eisenberg	IFVLWMAGCYPTSHENQDKR
Cornette	FILVMHYCWAQRTGESPNKD
Kyte	IVLFCMAGTSWYPHDNEQKR
Rose	CFIVMLWHYAGTSRPNDEQK
Sweet	FYILMVWCTAPSRHGKQNED

While the above scales differ in detail, there is a general consensus regarding the types of residue that appear at the most hydrophobic end (I, F, L, V and M) and those that appear at the most hydrophilic end (N, Q, E, D and K). Such scales can be used to plot hydropathy profiles, such as those illustrated in Figure 10.1. In creating such a profile, a sliding window is scanned across the query sequence and, for each window length, a score is calculated from the hydrophobicity values of each of the residues within the window. A typical graph will show characteristic peaks and troughs, corresponding to the most hydrophobic and most hydrophilic parts of the protein respectively. They are therefore of particular value in trying to identify possible membrane-spanning domains, which are characterised by consecutive runs of hydrophobic amino acids, usually 20–25 residues in length.

BOX 10.2: HELICAL WHEELS

Helical wheels are graphs containing five turns of helix. The graphs are usually plotted with α-helical periodicity (i.e., with 3.6 residues per turn), but different types of helix may be depicted by calculating the plots with the appropriate number of residues per turn (e.g., 3.0 for a 3_{10}-helix, 4.3 for a π-helix, and so on). Within a wheel, helical potential is recognised by the clustering of hydrophobic residues in non-polar arcs, or conversely of hydrophilic residues in polar arcs. As a tool for 'predicting' helices, wheels are relatively limited because not all helices show this kind of 'sidedness' in terms of the distribution of their hydrophobic and hydrophilic residues (i.e., not all helices are amphipathic).

and software, because there are different issues to be considered when buying a package, whether for company or academic use, as opposed to downloading freely available resources from the Internet.

10.3 Commercial databases

If the majority of biological databases are available from publicly accessible servers on the Internet, why bother with commercial databases? The answer lies in the industrial approach to information technology: i.e., the desire to purchase solutions to well-defined problems, rather than the more exploratory academic approach. In industry, if services exist to develop and maintain databases, and can be purchased, finance and manpower can then be released for the more exacting scientific task of searching and analysing the data. As only large, or highly specialised, organisations support substantial bioinformatics divisions, the use of scientists' time is carefully scrutinised.

Major releases of DNA and protein sequence databases occur every three to four months. In the meantime, newly determined sequences are added to daily-update files. To keep an in-house database up-to-date, synchronised FTP scripts are used (e.g., using scheduling software such as cron under UNIX). With such a system, it is relatively simple to track individual databases, but it becomes unwieldy when several databases (e.g., GenBank, EMBL, SWISS-PROT, PIR) have to be monitored and merged with proprietary information. Further, if new databases evolve, and it is considered advantageous also to bring them in-house, existing scripts must be updated to incorporate the new resources.

One answer to these problems is to find a good database service provider, who will either supply up-to-date databases or, alternatively, offer easily maintainable software for database updating.

10.4 Commercial software

Given that virtually all sequence comparison algorithms are published, and implementations of them are freely available, why go to the expense of purchasing commercial licences?

In an industrial environment, the requirement for software licences is often simply a matter of legal probity – although commercial releases from academic institutions are often identical to their freely available academic counterparts, it is reasonable that software used for commercial purposes should attract a fee. Sometimes public-domain code is provided with commercial suites in order to allow integration of a range of algorithms within the proprietary package framework (the GCG suite is an example where this approach seems to work especially well).

Finally, commercial organisations usually require some level of support to be assured. For example, they might want to know whether a technical helpline is available; whether and when new versions, with updated features, will be available; or whether the interface can be customised for use with a company **firewall** for access to the Internet; etc. These can be significant issues in a large company.

Another significant concern lies with company use of the Internet for database searching; the act of performing a database search with a proprietary sequence on a public server is tantamount to publication. Such a disclosure could prejudice a subsequent patent proposal, for example, and hence limit the subsequent utility of the sequence in a commercial environment. Such important security issues provide the incentive for companies to license packages whenever possible and bring them in-house.

10.5 Comprehensive packages

In the following sections, we outline a number of well-known packages that offer a fairly complete set of tools for both DNA and protein sequence analysis. These suites have evolved and grown to be fairly comprehensive over a period of years; other packages may compete in terms of coverage of tools in the coming years.

10.5.1 GCG

The most widely known, commercially available sequence analysis software is the GCG suite (Oxford Molecular Group). This was developed by the Genetics Computer Group at Wisconsin (575 Science Drive, Madison, Wisconsin, USA 53711) (Devereux et al., 1984), primarily as a set of analysis tools for nucleic acid sequences, but which in time included additional facilities for protein sequence analysis.

Within GCG, many of the frequently used sequence databases can be accessed (e.g., GenBank, EMBL, PIR and SWISS-PROT), as can a number of motif and specialist databases (such as PROSITE; TFD, the transcription factor database; and REBASE, the restriction enzyme database). A particular strength of the system is that it can also be relatively easily customised to accept additional, user-specific databases. Within the suite, EMBL and GenBank are split into different sections, allowing users to minimise search time by directing queries only to relevant parts of the databases. Thus, for example, sequences in

GenBank and EMBL may be searched either collectively, separately, or by defined taxonomic categories (e.g., viral, bacterial, rodent, etc.).

As we have seen, the sequence databases have their own distinct formats, so these must be converted to GCG format for use with its programs. Likewise, all data files imported to the suite for analysis must adhere to the GCG format. Many components of the package are thus simply format conversion scripts, to allow, for example, sequences from different databases to be used within the suite and/or, conversely, for GCG-format sequences to be exported and used within different programs or packages.

The complete list of facilities offered by GCG cannot be described in detail here. However, the options include tools for pairwise similarity searching, multiple sequence alignment, evolutionary analysis, motif and profile searching, RNA secondary structure prediction, hydropathy and antigenicity plots, translation, sequence assembly (based on Staden's method (see below)), restriction site mapping, and so on.

The initial implementation of GCG was on a mini-computer under the VMS operating system, and its **command-line** interface reflects this history. More recent developments have seen the provision of a windowing interface under UNIX, making the package easier to use. Use of all versions of the software, whether command-line, X-windows or Web-based, requires a licence, which for individuals may appear to be relatively expensive. Until recently, however, group licences have been made available at discretionary rates for organisations with extensive user communities. Thus, for example, EMBnet nodes, which offer remote login facilities to their respective national user bases, have been able to offer access to GCG, obviating the need for each user to obtain an independent licence. However, this situation may not prevail, and academic users may soon turn increasingly towards cheaper, more flexible and up-to-date, non-commercial alternatives.

From the point of view of protein sequence analysis, GCG tools are fairly primitive. For primary database searching, standard FastA and BLAST routines are offered, but for more incisive analysis and sensitive pattern recognition, facilities are only provided to search PROSITE and profiles.

EGCG

Extended GCG, or EGCG, began life at EMBL in Heidelberg as a collection of programs to support EMBL's research activities. Since 1988, through a collaboration of groups within EMBnet and elsewhere, further additions have been made to the suite in order to provide support both for core analysis activities at the Sanger Centre, and for the entire user base of EMBnet national nodes.

There are more than 70 programs in EGCG, covering themes such as fragment assembly, mapping, database searching, multiple sequence analysis, pattern recognition, nucleotide and protein sequence analysis, evolutionary analysis, and so on. The original collection of extension programs were merged and released with GCG 7.0 as 'unsupported' software for their licensed VMS users, and was subsequently ported to UNIX for distribution with GCG 7.2.

In 1997, various licensing issues arose that meant that it was no longer possible to distribute academic source code that used GCG libraries. As a result, EGCG 9.1, which was due for release in early 1998, was considered obsolete. In November 1997, the EGCG developers therefore set out to design a completely new generation of academic sequence analysis software, free from licensing complications (essentially by releasing the package under a version of the GNU General Public Licence). This effort has become known as the EMBOSS project (see Section 10.8.2).

10.5.2 Staden

The Staden Package is a set of tools for DNA and protein sequence analysis (Staden, 1988). It does not provide databases, but the software works with the EMBL database (as distributed by CD-ROM) and other databases in a similar format. The package has a windowing interface for UNIX workstations.

Amongst its range of options, the suite provides utilities to define and to search for patterns of motifs in proteins and nucleic acids (for example, specific individual routines allow searching for mRNA splice junctions, *E. coli* promoters, tRNA genes, etc., and users may define equally complex patterns of their own). Patterns may be defined in many different ways, and the search algorithms operate both on individual queries and on whole libraries of sequences.

A particular strength of the Staden Package lies in its support for DNA sequence assembly. It provides methods for all the pre-processing required for data from fluorescence-based sequencing instruments, including trace viewing (TREV), quality clipping (PREGAP4) and vector removal

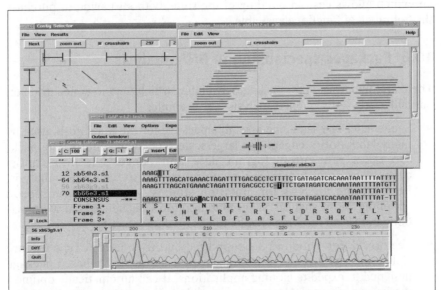

Figure 10.2 Screen dump illustrating some of the features of the graphical user interface to the Staden package. The figure shows the Contig Comparator containing a variety of search results, the Template Plot and Quality Plot, the Contig Editor, and the Traces windows.

(PREGAP4, VECTOR_CLIP); a range of assembly engines; and powerful contig. editing and finishing algorithms (GAP4). The most recent addition (Bonfield *et al.*, 1998) is a new method for detecting point mutations (TRACE_DIFF, GAP4). For analysis of finished DNA sequences, the package includes NIP4, and for comparing DNA or protein sequences, SIP4; these routines also provide an interface to the sequence libraries. The new interactive programs TREV, PREGAP4, GAP4, NIP4 and SIP4 have graphical user interfaces (e.g., see Figure 10.2), but the package also contains a large number of older, but still useful, programs that are text based.

The package is available on CD subject to completion of the necessary licence agreements. Further information is available at the Web address given at the end of the chapter.

10.5.3 Lasergene

Lasergene is a PC-based package that provides facilities for coding analysis, pattern and site matching, and RNA/DNA structure and composition analysis; restriction site analysis; PCR primer and probe design; sequence editing; sequence assembly and contig. management; multiple and pairwise sequence alignment (including dotplots); protein secondary structure prediction and hydropathy analysis; helical wheel and net creation; and database searching. The package allows use of the clipboard to export results to graphics or word-processing programs for publication purposes.

Lasergene is available for Windows or Macintosh, for single users or for networked-PC environments (but the provision of databases and their updates can be problematic on distributed networks). The package is modular, which allows custom building of the suite, without having to purchase the full range of functions available.

10.6 Packages specialising in DNA analysis

There are numerous other packages available, which tend to concentrate on particular areas of sequence analysis. For example:

- Sequencher is a sequence assembly package for the Macintosh, used by many laboratories engaged in large-scale sequencing efforts. The package takes raw chromatogram data and converts it into contig. assemblies; other functions include restriction site and ORF analysis, heterozygote analysis for mutation studies, vector and transposon screening, motif analysis, silent mutation tools, sequence quality estimation, and visual marking of edits to ensure data integrity.

- VectorNTI, for Windows 3.1 supported by the American Type Culture Collection (ATCC) and InforMax, Inc., is a knowledge-based package designed to expedite cloning applications. It can automatically optimise the design of new DNA constructs and recommend cloning steps. The user can specify preferences for processes such as fragment isolation, modification of termini, and ligation. The system incorporates ~3000 rules for genetic engineering.

- MacVector is a molecular biology system that exploits the Macintosh user interface to create an easy-to-use environment for manipulation and analysis of DNA and protein sequence data. The package implements the five BLAST search functions, and includes ClustalW for sequence alignment, and an icon-managed sequence editor that is integrated with the program's molecular biology functions (e.g., translation, restriction analysis, primer and probe analysis, protein structure prediction, and motif analysis). Facilities are also provided to compute predicted sequence-based melting curves for DNA and RNA structures.

There are, of course, many other commercial packages that specialise in DNA and protein sequence analysis, but it is beyond the scope of this book to deal with them all here. Further, more complete, information may be readily obtained by interrogating the World Wide Web.

10.7 Intranet packages

The future for commercial solutions lies in providers understanding the key issues facing the large industrial user. Most companies now have **intranets** and support the use of HTTP and **Internet Inter-ORB Protocol (IIOP)**. Bioinformatics solutions must fit as seamlessly as possible into this environment. Most companies need to implement integration throughout the research operation, and this is especially so in terms of genomic information.

Most industrial bioinformatics teams devote some resources to development and maintenance of internal Web servers that replicate the services available at public bioinformatics sites. It is frequently necessary to take this approach for security reasons, but also in order to enable integration of standard bioinformatics tools with in-house software, developed for specific industrial applications. Two companies, NetGenics Inc. and Pangea Systems Inc., provide bioinformatics systems that offer the prospect of service integration via the intranet.

10.7.1 SYNERGY

SYNERGY, developed by NetGenics Inc., Cleveland, Ohio, is an object-oriented approach using Java, CORBA, and an object-oriented database, to implement a flexible environment for managing bioinformatics projects. SYNERGY integrates standard tools into its portfolio through the use of CORBA 'wrappers', which present a streamlined interface between the tool and the SYNERGY system. In this way, the developers are able to incorporate a number of standard programs very rapidly, and users of the system are able to incorporate their own tools by implementing CORBA wrappers in-house.

The object-oriented learning curve is steep, however, and including CORBA adds a further layer of complexity. Although an elegant solution, the time may not be ripe for such leading-edge technology to be adopted in all industrial environments. Defining interfaces (i.e., specifications for wrappers) for objects in bioinformatics is under active discussion, and a Special Interest Group (SIG) of the Object Management Group (OMG) has been formed in the life sciences to address the issue. Standardising interfaces to programs and databases should make the deployment of object-oriented systems much more achievable in the future.

10.7.2 GeneMill, GeneWorld, GeneThesaurus

GeneMill, GeneWorld and GeneThesaurus are the developments of Pangea Systems Inc., Oakland, California. These are Web-based tools that are back-ended by a relational database. The overall system is aimed at high-throughput sequencing projects, and other large-scale industrial genomics projects, including, for example, GeneMill, a sequencing workflow database system for managing sequencing projects; GeneWorld, a tool for analysis of DNA and protein sequences; and GeneThesaurus, a sequence and annotation data subscription service, allowing access to public data and integration with proprietary data. The system is modular and allows interfaces to in-house software to be built seamlessly, using an open programming interface, PULSE (Pangea's Unified Life Sciences Environment).

As already mentioned, NetGenics' and Pangea's products have been developed for the intranet environment. We now briefly review some packages that are freely available over the Internet (but which could equally be implemented on corporate intranets).

10.8 Internet packages

10.8.1 CINEMA

As we have seen, central to sequence analysis is the multiple alignment. Consequently, a vital tool for the sequence analyst is an alignment editor. Several automatic alignment programs are now available, either in a stand-alone form (such as ClustalW) or as components of larger packages (such as Pileup in GCG). But automatically calculated alignments almost invariably require some degree of manual editing, whether to remove spurious gaps, to rescue residue widows, or to correct misalignments. This often presents problems, as there is currently no standard format for alignments. Consequently, swapping between alignment programs is almost impossible without the use of *ad hoc* scripts to convert between disparate input and output formats.

The advent of the object-oriented network programming language, Java (Gosling and McGilton, 1995), addresses some of these problems. Java-capable browsers may run **applets** on a variety of platforms – applets are small applications loaded from a server via HTML pages; the software is loaded on-the-fly from the server, and cached for that session by the browser.

CINEMA is a Colour Interactive Editor for Multiple Alignments, written in Java (Parry-Smith *et al.*, 1998): the program allows creation of sequence alignments by hand, generation of alignments automatically (e.g., using ClustalW), and visualisation and manipulation of sequence alignments currently resident at different sites on the Internet. In addition to its special advantage of allowing interactive alignment over the Web, CINEMA provides links to the primary data sources, thereby giving ready access to up-to-date data, and a gateway to related information on the Internet.

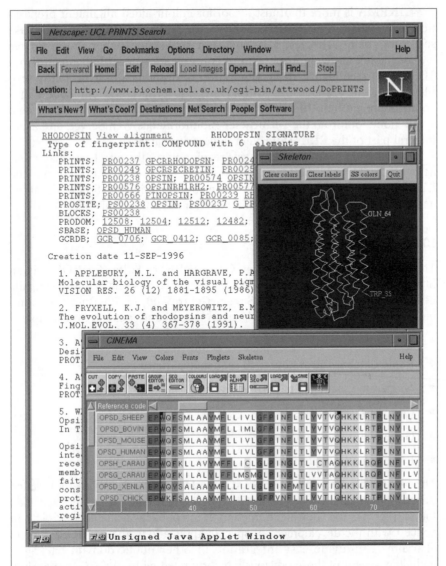

Figure 10.3 The CINEMA colour interactive editor for multiple alignments. The figure illustrates how conserved motifs may be highlighted and visualised in three dimensions (where co-ordinates are available). Here, the first transmembrane domain of the rhodopsin family of G-protein-coupled receptors has been selected from the alignment and displayed on the 3D model.

CINEMA is more than just a tool for colour-aided alignment preparation. The program also offers facilities for motif identification; database searching (using BLAST); 3D-structure visualisation (where co-ordinates are available), allowing inspection of conserved features of alignments in a 3D context (see Figure 10.3); generation of dotplots and hydropathy profiles; six-frame translation; and so on. The program is embedded in a comprehensive help file (written in HTML) and is accessible both as a stand-alone tool from the DbBrowser Bioinformatics Web Server, and as an integral part of the PRINTS protein fingerprint database.

Exploitation of such technologies revolutionises the way users may interact with databases in the future: bioinformatics centres need not just provide data, but are now able to offer the means by which information is visualised and manipulated, without the requirement for users to install code.

10.8.2 EMBOSS

The European Molecular Biology Open Software Suite (EMBOSS) is an integrated set of packages and tools for sequence analysis being specifically developed for the needs of the Sanger Centre and the EMBnet user communities. Applications of the package are planned to include, for example:

- EST clustering

- Rapid database searching with sequence patterns

- Nucleotide sequence pattern analysis

- Codon usage analysis

- Gene identification tools

- Protein motif identification

The first application of EMBOSS was a file format conversion tool, called 'seqret', which simply reads an input sequence and writes it out to a specified format (i.e., it is the equivalent function to the publicly available 'readseq'). In concert with the development of specific applications, the EMBOSS project also aims to provide easy integration of other public-domain packages, including several of those previously incorporated within EGCG, but excluding any for which there is no GCG-free version of source code available.

10.8.3 Alfresco

Alfresco is a visualisation tool that is being developed for comparative genome analysis, using ACEDB for data storage and retrieval. The program compares multiple sequences from similar regions in different species, and allows visualisation of results from existing analysis programs, including those for gene prediction, similarity searching, **regulatory sequence** prediction, etc.

By using available analysis programs relevant to comparative genome sequence analysis, the developers are free to focus on designing an intuitive graphical user interface for combining the results of the different applica-

The range and scope of analysis packages available, and the variety of analysis tools they each offer, are considerable and we cannot do proper justice to them here. Nevertheless, we hope to have given a flavour of some of today's most popular packages. For more up-to-date information on sequence analysis packages we would recommend querying the World Wide Web via one of the standard search engines (AltaVista, LookSmart, Infoseek, NetSearch, etc.).

10.9 Summary

- Because of the need to employ a range of techniques for sequence analysis, software packages have been developed to bring several methods together under a single umbrella. This removes the burden of learning how to use different tools, with different interfaces, with different input requirements and different output formats.

- Although most biological databases are in the public domain, there is a market for commercial databases. Companies usually wish to purchase complete solutions to well-defined problems – e.g., if services to develop and maintain databases can be bought, manpower can be released to support research.

- Although most sequence analysis software is in the public domain, for various reasons, companies tend to prefer to purchase commercial licenses – e.g., because of the need for technical support, or the desire to keep information behind company firewalls.

- There are several comprehensive packages for sequence analysis available. These include the GCG suite (which can be customised to accept user-specific databases), the Staden package (whose particular strength is in support for DNA sequence assembly), and the PC-based Lasergene (a modular package that allows custom building of the suite).

- Other packages tend to concentrate on particular areas of sequence analysis, including Sequencher (for large-scale sequencing), VectorNTI (designed to expedite cloning applications), and MacVector (exploits the Macintosh interface to create an easy-to-use environment for DNA and protein sequence analysis).

- As most companies now have intranets, several bioinformatics applications have been designed to offer service integration via the intranet – e.g., SYNERGY is an object-oriented approach using Java, CORBA and an object-oriented database; and GeneMill, GeneWorld and GeneThesaurus are Web-based tools that exploit a relational database.

- Some packages are freely available over the Internet – e.g., CINEMA is a Java alignment editor that integrates facilities for database searching, motif recognition, structure visualisation, etc.; EMBOSS is an integrated set of tools for sequence analysis being developed at the Sanger Centre; and Alfresco is a visualisation tool that is being developed for comparative genome analysis.

10.10 Further reading

Sequence analysis tools
BONFIELD, J.K., RADA, C. and STADEN, R. (1998) Automated detection of point mutations using fluorescent sequence trace subtraction. *Nucleic Acids Research*, **26**, 3404–3409.

DEVEREUX, J., HAEBERLI, P. and SMITHIES, O. (1984) A comprehensive set of sequence-analysis programs for the VAX. *Nucleic Acids Research*, **12**(1), 387–395.

PARRY-SMITH, D.J., PAYNE, A.W.R., MICHIE, A.D. and ATTWOOD, T.K. (1998) CINEMA – A novel Colour INteractive Editor for Multiple Alignments. *Gene*, **211**(2), GC45–56.

STADEN, R. (1988) Methods to define and locate patterns of motifs in sequences. *Computer Applications in the Biosciences*, **4**(1), 53–60.

Hydropathy indices
TRINQUIER, G. and SANEJOUAND, Y.–H. (1998) Which effective property of amino acids is best preserved by the genetic code? *Protein Engineering*, **11**(3), 153–169.

Java
GOSLING, J. and McGILTON, H. (1995) *The Java Language Environment.* Sun Microsystems White Paper, http://java.sun.com/whitePaper/java-whitepaper-1.html

10.11 Web addresses

GCG	http://www.gcg.com/
ECGC	http://www.sanger.ac.uk/Software /EGCG/
Staden	http://www.mrc-lmb.cam.ac.uk/pubseq/
NetGenics	http://www.netgenics.com/
Pangea Systems	http://www.pangeasystems.com/
CINEMA	http://www.biochem.ucl.ac.uk/bsm/dbbrowser/CINEMA2.1/
EMBOSS	http://www.sanger.ac.uk/Software/EMBOSS/
Alfresco	http://www.sanger.ac.uk/Users/nic/alfresco.html

Glossary

Accession number: A unique number or code given to mark the entry of a sequence (protein or nucleic acid) or pattern (regular expression, finger-print, profile) to a primary or secondary database. Accession numbers should remain static between database updates, and hence in theory provide a mechanism for reliably identifying a particular entry in subsequent database releases.

Algorithm: The logical sequence of steps by which a task can be performed.

Alternatively spliced form: See Splice variant.

Amino acid: The fundamental building block of proteins. There are 20 naturally occurring amino acids in animals and around 100 more found only in plants.

Amphipathic helix: A helix that displays a characteristic charge separation in terms of the distribution of its polar and non-polar residues on opposite faces. Their 'sidedness' allows such helices to sit comfortably at polar/apolar interfaces, such as at the surfaces of globular proteins (where their hydrophilic sides point towards the solvent, and their hydrophobic sides point towards the protein core), or within membranes (where their hydrophobic sides point towards the lipid environment, and their hydrophilic sides point towards the protein interior).

Analogues: Non-homologous proteins that have similar folding architectures, or similar functional sites, which are believed to have arisen through convergent evolution.

Applet: Small software applications loaded from a server via HTML pages.

Assembly: The process of aligning overlapping sequence fragments into a contig. or series of contigs.

Basepair (bp): Any possible pairing between bases in opposing strands of DNA or RNA. Adenine pairs with thymine in DNA, or with uracil in RNA; and guanine pairs with cytosine.

Bioinformatics: The application of computational techniques to the management and analysis of biological information.

Block: An ungapped, aligned motif consisting of sequence segments that are clustered to reduce multiple contributions from groups of highly similar or identical sequences.

Browser: A computer program (commonly known as a Web client) that permits information retrieval from the Internet and the WWW.

cDNA library: A gene library composed of cDNA inserts synthesised from mRNA using reverse transcriptase.

Central dogma: A fundamental principle of molecular biology, first expounded by Francis Crick in 1958, essentially stating that the transfer of information from nucleic acid to nucleic acid, or from nucleic acid to protein, is possible, while transfer from protein to nucleic acid or from protein to protein is impossible. A shorthand expression of the dogma gives the unidirectional relation: DNA > RNA > protein.

Chaperone: A protein that assists the correct non-covalent assembly of folding proteins *in vivo;* chaperones do not themselves form part of the structures they help to assemble.

Chromosomes: The paired, self-replicating genetic structures of cells that contain the cellular DNA; the nucleotide sequence of the DNA encodes the linear array of genes.

Client: Any program that interacts with a server (Lynx, Mosaic and Netscape are examples of client software).

Clone: A copied fragment of DNA, maintained in circular form, identical to the template from which it is derived.

Cloning: The process of generating identical copies of a DNA fragment (that may encode a complete gene) from a single template DNA.

Cloning vector: A DNA molecule originating from a virus, a plasmid, or the cell of a higher organism into which another DNA fragment can be integrated without compromising the vector's capacity for self-replication.

Coding sequence (CDS): A region of DNA or RNA whose sequence determines the sequence of amino acids in a protein.

Command line: The basic level at which a computer prompts the user for input.

Communication protocol: An agreed set of rules for structuring communication between programs (allowing, for example, data exchange between nodes on the Internet).

Complementary DNA (cDNA): DNA that is synthesised from a messenger RNA template using the enzyme reverse transcriptase.

Composite database: A database that amalgamates a number of primary sources, using a set of defined criteria that determine the priority of inclusion of the different sources and the level of redundancy retained (e.g., NRDB is a non-identical composite protein sequence database and OWL is a non-redundant composite).

Conceptual translation: The computational process of interpreting the sequence of nucleotides in mRNA via the genetic code to a sequence of amino acids, which may or may not code for protein.

Consensus sequence: A pseudo-sequence that summarises the residue information contained in a multiple alignment.

Conserved sequence: A sequence of bases in a DNA molecule (or an amino acid sequence in a protein) that has remained essentially unchanged during evolution.

Contig.: Sequences of clones, representing overlapping regions of a gene, presented as an assembly or multiple alignment.

Dansylation: A method used to add dansyl groups to free amino groups in protein end-group analysis. The dansyl amino acids, isolated after hydrolysis of the protein, are fluorescent and may be detected in nanomolar quantities.

Diagnostic performance (diagnostic power): A measure of the ability of a discriminator to identify true matches, either in an individual query sequence or in a database.

Discriminator: A mathematical abstraction of a conserved motif, or set of motifs (e.g., a regular expression pattern, a profile or a fingerprint), used to search either an individual query sequence or a full database for the occurrence of that same, or similar, motif(s).

DNA (deoxyribonucleic acid): The molecule that encodes genetic information. DNA is a double-stranded molecule held together by weak bonds between basepairs of nucleotides. The four nucleotides in DNA contain the bases: adenine (A), guanine (G), cytosine (C) and thymine (T). In nature, basepairs form only between A and T and between G and C; thus the base sequence of each single strand can be deduced from that of its partner.

DNA sequence: The linear sequence of base pairs, whether in a fragment of DNA, a gene, a chromosome or an entire genome.

Domain: A compact, local, semi-independent folding unit, presumed to have arisen via gene fusion and gene duplication events. Domains need not be formed from contiguous regions of an amino acid sequence: they may be discrete entities, joined only by a flexible linking region of the chain; they may have extensive interfaces, sharing many close contacts; and they may exchange chains with domain neighbours. The combination of domains within a protein determines its overall structure and function.

Down: State of a computer when it is non-operational and hence unavailable for normal use.

Dumb: A dumb terminal is a desktop display device that is not capable of local processing, this being entirely carried out by the central computer. Such terminals (e.g., VT52, VT100, etc.) do not support windowing applications.

E.C. system: The systematic classification and naming of enzymes by the Enzyme Commission, whereby enzymes are denoted by the letters E.C., followed by a set of four numbers separated by dots. The first number indicates one of six main functional divisions (oxidoreductases, transferases, hydrolases, lyases, isomerases and ligases); the following numbers denote different subclasses, as defined by donor group, acceptor, substrate, isomer, etc., the final digit being a serial number for the particular enzyme (e.g., E.C.1.1.1.1 for alcohol dehydrogenase, E.C.3.5.3.15 for protein-arginine deiminase, etc.).

Edman degradation: A method used in sequencing polypeptides, whereby amino acid residues are removed sequentially from the N-terminus by reaction with phenyl-isothiocyanate, to form phenylthiocarbamyl-peptide (PTC-peptide). This is cleaved in anhydrous acid, releasing a thiazolinone intermediate and the remainder of the peptide.

Enzyme: A protein that acts as a catalyst, speeding the rate at which a biochemical reaction proceeds but not altering the direction or nature of the reaction.

Enzyme Classification System: See E.C. system.

Eukaryote: Cell or organism with membrane-bound, structurally discrete nucleus and other well-developed subcellular compartments. Eukaryotes include all organisms except viruses, bacteria and blue-green algae.

Exons: The protein-coding DNA sequences of a gene.

Expressed Sequence Tag (EST): A partial sequence of a clone, randomly selected from a cDNA library and used to identify genes expressed in a particular tissue. ESTs are used extensively in projects to map the human genome.

Expression profile: The characteristic range of genes expressed at different stages of a cell's development and functioning.

False-negative: A true match that incorrectly fails to be recognised by a discriminator.

False-positive: A false match incorrectly recognised by a discriminator.

File Transfer Protocol (FTP): A method of transferring files to remote computers.

Fingerprint: A group of ungapped motifs excised from a sequence alignment and used to build a characteristic signature of family membership by means of iterative searching of a primary (or composite) database.

Firewall: A mechanism for protecting a proprietary computer network (or intranet), allowing internal users to access the Internet, while preventing external Internet users from penetrating the intranet.

Flat-file: A human-readable data-file in a convenient form for interchange of database information. Flat-files may be created as output from relational databases, in a format suitable for loading into other databases.

Folding problem: The problem of determining how a protein folds into its final 3D form given only the information encoded in its primary structure.

Frameshift: An alteration in the reading sense of DNA resulting from an inserted or deleted base, such that the reading frame for all subsequent codons is shifted with respect to the number of changes made (e.g., if a sequence should read UCU-CAA-AGG-UUA, and a single U is added to the beginning, the new sequence would read UUC-UCA-AAG-GUU, etc.). Frameshifts may arise through random mutations, or via errors in reading sequencing output.

Gene: The fundamental physical and functional unit of heredity. A gene is an ordered sequence of nucleotides located in a particular position on a particular chromosome that encodes a specific functional product (i.e., a protein or RNA molecule).

Gene cloning: See Cloning.

Gene duplication: A genetic alteration in which a segment of DNA is repeated. Duplications may appear anywhere, but where the duplicated segment is adjacent to the original one, this is termed a tandem duplication.

Gene expression: The process by which a gene's coded information is converted into the structures present and operating in the cell. Expressed genes include those that are transcribed into mRNA and then translated into protein and those that are transcribed into RNA but not translated into protein (e.g., transfer and ribosomal RNAs).

Gene families: Groups of closely related genes that encode similar protein products.

Gene product: The protein resulting from the expression of a gene. In some cases, the gene product may be an RNA molecule that is never translated.

Genetic code: The rules that relate the four DNA or RNA bases to the 20 amino acids. There are 64 possible three-base (triplet) sequences, which are known as codons. A single triplet uniquely defines one amino acid, but an amino acid may be coded by as many as six codons. The code is thus said to be degenerate.

Genome: All the genetic material in the chromosomes of a particular organism; its size is generally given as its total number of basepairs.

Genome projects: Initiatives (often via international collaboration) to map and sequence the entire genomes of particular organisms. The first complete eukaryotic genome to have been sequenced is that of the yeast *S. cerevisiae;* the human genome is expected to be finished by roughly 2003–2005; and mouse by around 2008. The majority of genomes completed to date are those of prokaryotes.

Helical wheel: A circular graph depicting five turns of helix, around which the residues of a protein sequence are plotted. Helical potential is recognised by the clustering of hydrophilic and hydrophobic residues in distinct polar and non-polar arcs.

Hidden Markov Model (HMM): A probabilistic model consisting of a number of interconnecting states. Like profiles, HMMs encode full domain alignments. They are essentially linear chains of match, delete or insert states: a match state denotes a conserved column in an alignment; an insert state allows insertions relative to match states; and delete states allow match positions to be skipped.

Home page: The HTML document that acts as the first contact point between a browser and a server.

Homology: Being related by the evolutionary process of divergence from a common ancestor. Homology is not a synonym for similarity.

Hybridisation: The process of joining two complementary strands of DNA or one each of DNA and RNA to form a double-stranded molecule.

Hydropathy: Having the property of hydrophobicity, a low affinity for water.

Hydropathy profile: A graph in which hydropathy values are calculated within a sliding window and plotted for each residue in a protein sequence. Such graphs show characteristic peaks and troughs, corresponding to the most hydrophobic and hydrophilic regions of the sequence respectively.

Hydrophobicity: See Hydropathy.

Hyperlink: An active HTTP cross-reference that links one Web document to another document on the Internet.

Hypermedia: Formatted Web documents containing a variety of information types, including text, image, movie and audio.

Hypertext: Text that contains embedded links (hyperlinks) to other documents.

HyperText Markup Language (HTML): The syntax governing the way documents are created so that they can be interpreted and rendered by Web browsers.

HyperText Transport Protocol (HTTP): The communication protocol used by Web servers.

INDEL: An INsertion/DELetion in a DNA or protein sequence.

Internet: The international network of computer networks that connect government, academic and business institutions.

Internet Inter-ORB Protocol (IIOP): The communication protocol used by object-request brokers to communicate over the Internet.

Intranet: Computer network isolated from the Internet by means of a firewall but that offers similar facilities to the local community (e.g., Web servers, mail, etc.).

Introns: The sequence of DNA bases that interrupts the protein-coding sequence of a gene; these sequences are transcribed into RNA but are edited out of the message before it is translated into protein.

IP address: Internet Protocol address – a unique identifying number assigned to each computer on the Internet to allow communication between them.

Java: An object-oriented, network programming language that permits creation of either stand-alone programs, or applets that are launched via links on Web pages. In theory, Java programs run on any machine that supports the Java run-time environment (including PCs and UNIX workstations).

Kilobase (Kb): Unit of length for DNA fragments equal to 1000 nucleotides.

Library: An unordered collection of clones (i.e., cloned DNA from a particular organism), generated from genomic DNA or cDNA.

Locus (pl. loci): The position on a chromosome of a gene or other chromosome marker; also, the DNA at that position. The use of locus is sometimes restricted to mean regions of DNA that are expressed.

Megabase (Mb): Unit of length for DNA fragments equal to 1 million nucleotides.

Midnight Zone: Region of sequence identity where sequence comparisons fail completely to detect structural similarity.

Model system: A biological system used to represent other, often more complex, systems, in which similar phenomena either do, or are thought to, occur (e.g., *D. melanogaster, M. musculus, S. cerevisiae, C. elegans, E. coli*).

Module: An autonomous folding unit, believed to have arisen largely as a result of genetic shuffling mechanisms. Modules are contiguous in sequence and are often used as building blocks to confer a variety of complex functions on the parent protein. They may be thought of as a subset of protein domains. Examples of modules include Kringle domains (named after the shape of a Danish pastry), which are autonomous structural units found throughout the blood clotting and fibrinolytic proteins; the ubiquitous DNA-binding zinc fingers, which are small self-folding units in which zinc is a crucial structural component; and the WW module (characterised by two conserved tryptophan residues, hence its name), which is found in a number of disparate proteins, including dystrophin, the product encoded by the gene responsible for Duchenne muscular dystrophy.

Mosaic: A mosaic protein is a modular protein that, rather than including multiple tandem repeats of the same module, is composed of a number of different modules, each conferring different aspects of the parent protein's overall functionality (e.g., the calcium independent latrotoxin receptor, a mosaic of EGF-like and laminin G-like modules).

Motif: A consecutive string of amino acids in a protein sequence whose general character is repeated, or conserved, in all sequences in a multiple alignment at a particular position. Motifs are of interest because they may correspond to structural or functional elements within the sequences they characterise.

Multiple alignment: See Sequence alignment.

Mutation: Any change in DNA sequence.

Normalised library: cDNA library generated such that all the genes in the library are represented at the same frequency.

Nucleotide: A molecule consisting of a nitrogenous base (A, G, T or C in DNA; A, G, U or C in RNA), a phosphate moiety and a sugar group (deoxyribose in DNA and ribose in RNA). Thousands of nucleotides are linked to form a DNA or RNA molecule.

Object-oriented database: A database in which data are stored as abstract objects, with abstract relationships between them. The data representations are potentially very varied, including, for example, character strings, digitised images, tables, etc.. An object may subsume many other objects, and the database allows retrieval of the objects as a whole. The flexibility of data representation, and the ability to group objects together, renders object-oriented databases potentially very powerful systems.

Open reading frame (ORF): A series of DNA codons, including a 5' initiation codon and a termination codon, that encodes a putative or known gene.

Operating system: A program, or suite of programs, that controls the entire operation of the computer, handling input/output operations, interrupts, user requests, etc. (e.g., UNIX, VMS, Windows NT, etc.).

Orthologues: Homologous proteins that perform the same function in different species.

Packet: A self-contained message, or component of a message, comprising address, control and data signals, which may be transferred as a single entity within a communications network.

Paralogues: Homologous proteins that perform different but related functions within one organism.

Pattern: See Regular expression.

Pattern database: See Secondary database.

Penalties: Scores, or weights, used by programs in the computation of sequence alignments; such scores are normally supplied as parameters to the programs and thus may be modified by the user.

Phantom INDELs: Spurious insertions or deletions that arise when physical irregularities in a sequencing gel cause the reading software either to call a base too soon, or to miss a base altogether.

Phylogenetic analysis: Study of the evolutionary relationships between a species and its predecessors (e.g., using phylogenetic trees).

Phylogenetic tree: A graphical representation of the putative evolutionary relationships between groups of organisms, e.g. as calculated from multiple protein or nucleic acid sequence alignments.

Polymerase chain reaction (PCR): A method for amplifying a DNA base sequence using a heat-stable polymerase and two primers, one complementary to the (+)-strand at one end of the sequence to be amplified and the other complementary to the (−)-strand at the other end. The faithfulness of reproduction of the sequence is related to the fidelity of the polymerase. Errors may be introduced into the sequence using this method of amplification.

Post-translational modification: An enzyme-catalysed alteration to a protein made after its translation from mRNA (e.g., glycosylation, phosphorylation, myristoylation, methylation).

Primary database: A database that stores biomolecular sequences (protein or nucleic acid) and associated annotation information (organism, species, function, mutations linked to particular diseases, functional/structural patterns, bibliographic, etc.).

Primary structure: The linear sequence of amino acids in a protein molecule.

Primer: A short polynucleotide chain to which new deoxyribonucleotides can be added by DNA polymerase.

Probe: A DNA or protein sequence used as a query in a database search.

Profile: A position-specific scoring table that encapsulates the sequence information within complete alignments. Profiles define which residues are allowed at given positions; which positions are conserved and which degenerate; and which positions, or regions, can tolerate insertions. In addition to data implicit in the alignment, the scoring system may include evolutionary weights and results from structural studies. Variable penalties are specified to weight against insertions and deletions occurring in secondary structure elements.

Prokaryote: An organism lacking a membrane-bound, structurally discrete nucleus and other subcellular compartments. Bacteria are prokaryotes.

Promoter: A site on DNA to which RNA polymerase will bind and initiate transcription.

Protein: A molecule composed of one or more chains of amino acids in a specific order; the order is determined by the base sequence of nucleotides in the gene coding for the protein. Proteins are required for the structure, function and regulation of cells, tissues and organs, each protein having a specific role (e.g., hormones, enzymes and antibodies).

Quaternary structure: The arrangement of separate protein chains in a protein molecule with more than one subunit.

Quinternary structure: The arrangement of separate molecules, such as in protein–protein or protein–nucleic acid interactions.

R-factor: In X-ray crystallography, this parameter is used to express the extent of agreement between theoretical calculations and the measured data; the lower the R-factor, the better the fit (R means either Residual or Reliability).

Regular expression: A single consensus expression derived from a conserved region of a sequence alignment, and used as a characteristic signature of family membership. Synonymous terms: rule, pattern.

Regulatory regions or **sequences:** A DNA base sequence that controls gene expression.

Relational database: A database that uses a relational data model, in which data are stored in two-dimensional tables. The tables embody different aspects or properties of the data, but contain overlapping information.

Resolution: The extent to which closely juxtaposed objects can be distinguished as separate entities. The degree of resolution is dependent on the resolving power of the system; the fineness of detail with which objects may be visualised is determined by the wavelength of electromagnetic radiation used. X-rays, for example, have wavelengths in the range 10^{-8}m to 10^{-11}m and hence can be used to resolve structures at the atomic level. Structures are thus said to be determined, for example, to 3 Å resolution, 5 Å resolution, etc.

RNA (ribonucleic acid): A molecule chemically similar to DNA that plays a central role in protein synthesis. The structure of RNA is similar to that of DNA but it is inherently less stable. There are several classes of RNA molecule, including messenger RNA (mRNA), transfer RNA (tRNA), ribosomal RNA (rRNA), and other small RNAs, each serving a different purpose.

Rule: A short regular expression (typically 4–6 residues in length) used to identify generic (non-family specific) patterns in protein sequences. Rules tend to be used to encode particular functional sites: e.g., sugar attachment sites, phosphorylation, hydroxylation, sulphation sites, etc. However, their small size means that the patterns do not provide good discrimination, and can only give a guide as to whether a certain functional site *might* exist in a sequence.

Secondary database: A database that contains information derived from primary sequence data, typically in the form of regular expressions (patterns), fingerprints, blocks, profiles or Hidden Markov Models. These abstractions represent distillations of the most conserved features of multiple alignments, such that they are able to provide potent discriminators of family membership for newly determined sequences.

Secondary structure: Regions of local regularity within a protein fold (e.g., α-helices, β-turns, β-strands).

Sequence alignment: A linear comparison of amino (or nucleic) acid sequences in which insertions are made in order to bring equivalent positions in adjacent sequences into the correct register. Alignments are the basis of sequence analysis methods, and are used to pinpoint the occurrence of conserved motifs.

Sequence Tagged Site (STS): Short (200 to 500 basepairs) DNA sequence that has a single occurrence in the human genome and whose location and base sequence are known. Detectable by polymerase chain reaction (PCR), STSs are useful for localising and orienting the mapping and sequence data reported from many different laboratories and serve as landmarks on the developing physical map of the human genome. Expressed sequence tags (ESTs) are STSs derived from cDNAs.

Sequencing: Determination of the order of nucleotides (base sequences) in a DNA or RNA molecule, or the order of amino acids in a protein.

Server: A computer or software system that communicates information via the Internet to a client.

Shotgun method: Cloning of DNA fragments randomly generated from a genome.

Silent mutation: A nucleotide substitution that does not result in an amino acid substitution in the translation product, because of the redundancy of the genetic code.

Six-frame translation: Translation of a stretch of DNA taking into account three forward translations and three reverse translations, arising from the three possible reading frames of an uncharacterised stretch of DNA.

Sparse matrix: A matrix in which most of the elements or cells have zero scores.

Splice variants: Proteins of different length that arise through translation of mRNAs that have not included all available exons in the template DNA.

Subject: A DNA or protein sequence matched by a query sequence in a database search.

Subunit: A distinct polypeptide chain within a protein that may be separated from other chains (whether identical or different) without breaking covalent bonds.

Super-secondary structure: The arrangement of α-helices and/or β-strands in a protein sequence into discrete folded structures (e.g., β-barrels, β-α-β units, Greek keys, etc.).

Telnet protocol: A method of communication between remote computers that allows users to log on and use the distant machines as if physically present at the remote location.

Tertiary database: A database derived from information housed in secondary (pattern) databases (e.g., the BLOCKS and eMOTIF databases, which draw on data stored within PROSITE and PRINTS). The value of such resources is in providing a different scoring perspective on the same underlying data, allowing the possibility to diagnose relationships that might be missed using the original implementation.

Tertiary structure: The overall fold of a protein sequence, formed by the packing of its secondary and/or super-secondary structure elements.

Transcription: The synthesis of an RNA copy from a sequence of DNA (a gene); the first step in gene expression.

Translation: The process in which the genetic code carried by mRNA directs the synthesis of proteins from amino acids.

Transmembrane domain: A region of a protein sequence that traverses a membrane; for α-helical structures, this requires a span of 20–25 residues.

Transmission Control Protocol/Internet Protocol (TCP/IP): The rules that govern data transmission between two computers over the Internet.

True-negative: A false match that correctly fails to be recognised by a discriminator.

True-positive: A true match correctly recognised by a discriminator.

Twilight Zone: A zone of sequence similarity (~0–20% identity) within which alignments appear plausible to the eye but are not statistically significant (i.e., could have arisen by chance).

Uniform Resource Locator (URL): The address of a source of information. The URL comprises four parts – the protocol, the host name, the directory path and the file name (e.g., http://www.biochem.ucl.ac.uk/bsm/dbbrowser /prefacefrm.html).

Up: The status of a computer system when it is operational.

Upstream: Further back in the sequence of a DNA molecule, with respect to the direction in which the sequence is being read.

Weight matrix: See Profile.

Widow: Amino acid residues isolated from neighbouring residues by spurious gaps, usually the result of over-zealous gap insertion by automatic alignment programs.

World Wide Web: The information system or network on the Internet that uses HTTP as the primary communications medium.

Index